PRAISE FOR
Climate Resilience

"*Climate Resilience* is a crucial book that is much needed today as it starts shifting the focus to center frontline communities and leaders addressing climate change and fighting for climate justice."

—**JESSICA HERNANDEZ**, author of *Fresh Banana Leaves*

"This collection of essays is a profound gift to a world that is deeply in need of healing and restoration. These wise voices tell us important truths about the past, describing with unflinching honesty how the intertwined forces of settler colonialism, white supremacy, and capitalism have gotten us where we are today. Yet they also point to possibilities of a future that could be different—a future where we remember that our own well-being is intimately connected to the well-being of all other living creatures, and act from this remembrance. Readers will be inspired by the stories of the people already planting the seeds of this future in brave, creative, generative ways. *Climate Resilience* is also profoundly practical, with clear, actionable ideas and reflection prompts following the essays. This is a book that will help you find your place in a movement towards a world of just interdependence, deep connection, radical generosity, and climate resilience for all."

—**ELIZABETH BECHARD**, senior policy analyst for Moms Clean Air Force and author of *Parenting in a Changing Climate*

"*Climate Resilience* brought me to tears of recognition, relief, solidarity, and hope. It's the book we need to survive this human moment together."

—**LILY DIAMOND**, author of *Kale & Caramel* and coauthor of *What's Your Story?* with Rebecca Walker

"If you're interested in understanding how we might really tackle the problem of climate change, a problem that doesn't exist in a vacuum but one that is wrapped up in our social and political relationships, put down whatever you are reading and pick up this book. Kylie Flanagan has masterfully woven together conversations rooted in deep wisdom, love, and care, conversations which invite us to rethink the very nature of the problems we are facing and help us to envision a beautiful, more inclusive, more bountiful future. This book is a refreshing, much-needed addition to the overly technocratic, market-based, depoliticized literature on the climate crisis, as it recenters the problem as well as the source of potential solutions, anchoring them in the community-based expertise that is too often overlooked. The voices Kylie has channeled in this book give me hope that a better future is possible, and the path to that future is a beautiful one."

—**KHALID KADIR,** PhD, winner of the 2020 Constellation Prize for Engineering Education and lecturer in civil and environmental engineering at UC Berkeley

"There's no techno-fix for a cultural and spiritual crisis. The climate emergency needs solutions that bring liberation and restoration—and those sorts of solutions can only come from the margins. In *Climate Resilience*, Kylie Flanagan looks to the grassroots for climate leadership that is rooted in land, community and care. This is a book to inspire and educate, and also a book that invites our participation, with practical advice for applying solutions in our own contexts. If you enjoyed the fierce joy of *All We Can Save* and were wondering if there's more where that came from, read this next!"

—**JEREMY WILLIAMS,** author of *Climate Change Is Racist*

climate resilience

How We Keep Each Other Safe, Care for Our Communities, and Fight Back against Climate Change

in conversation with 39 women, nonbinary, and gender-expansive climate leaders

kylie flanagan

North Atlantic Books
Huichin, unceded Ohlone land
Berkeley, California

Published by
North Atlantic Books
Huichin, unceded Ohlone land
Berkeley, California

Cover art © gettyimages.com/saemilee
Cover design by Jess Morphew
Book design by Happenstance Type-O-Rama

Printed in the United States of America

Climate Resilience: How We Keep Each Other Safe, Care for Our Communities, and Fight Back against Climate Change is sponsored and published by North Atlantic Books, an educational nonprofit based in the unceded Ohlone land Huichin (Berkeley, CA) that collaborates with partners to develop cross-cultural perspectives, nurture holistic views of art, science, the humanities, and healing, and seed personal and global transformation by publishing work on the relationship of body, spirit, and nature.

North Atlantic Books's publications are distributed to the US trade and internationally by Penguin Random House Publisher Services. For further information, visit our website at www.northatlanticbooks.com.

Cataloging-in-Publication Data available from the Library of Congress

1 2 3 4 5 6 7 8 9 KPC 28 27 26 25 24 23

This book includes recycled material and material from well-managed forests. North Atlantic Books is committed to the protection of our environment. We print on recycled paper whenever possible and partner with printers who strive to use environmentally responsible practices.

CONTENTS

INTRODUCTION

When This Project Began

In the summer of 2020, California was ablaze with fire after fire, each seemingly more staggering in its devastation than the last. Over four million acres burned, and thick smoke hung heavy for months. In the San Francisco Bay Area, which is notoriously chilly and blanketed in fog from June through August, temperatures surged past record highs and lingered in the triple digits for days at a time.

I have a particularly poignant memory of running errands in San Francisco on a Sunday in September. After hiding out in my apartment from the surging pandemic and smoke for weeks in an attempt not to inflame my chronic asthma, I found that the city's parks were packed with thousands of people seeking relief from the suffocating heat in their homes. I looked down at my car's dashboard and the temperature read 103°F. I glanced at the Purple Air app on my phone, which I'd come to check dozens of times each day, and the air quality index hovered near 300, considered to be very unhealthy for the general population. As my own lungs burned, I recalled reading that breathing in air with an AQI of 300 for a day was roughly equivalent to smoking fourteen cigarettes.

Certainly, seeing the residents of San Francisco gathered under trees and near the water on this sweltering day should not have been a surprise. After all, the majority of San Francisco's housing lacks air-conditioning, having been built for the city's typically moderate coastal climate. Meanwhile, many of the cool refuges that folks typically seek out during extreme weather, like libraries and museums, were closed to prevent the spread of Covid-19. Even those with air-conditioning in their homes were out of luck; the region's investor-owned power supplier, Pacific Gas and Electric, was implementing sporadic power shutoffs to minimize the possibility of electric lines sparking fire amid the extreme conditions (and therefore mitigating their own potential liability).

And still, it was a poignant visualization of the impossible choice that San Franciscans were faced with: to endure the smoke outside, risk heat exhaustion at home, or seek relief somewhere air-conditioned indoors and potentially be exposed to Covid-19. (Notably, for many of the region's residents, who were unhoused or incarcerated or conducting work deemed essential, there was no choice to be made between heat, smoke, and virus.) It was a stark reminder of what it can look like when climate-related weather events compound upon one another and cruelly collide with the crises that don't burn as bright but are constantly simmering, like woefully inadequate energy and community infrastructure, housing, and healthcare.

This project was born during a fiery summer, a pandemic, racial uprisings, and myriad other ongoing calamities, in a moment when it felt positively palpable that we do not have time to focus on a single issue at a time but must build community resilience at the intersection of climate change mitigation, adaptation, and justice all at once. I sought the wisdom of leaders around the so-called[1] United States who have always known this to be true and have been doing this work in a deeply holistic, intersectional, and intentional way for years, oftentimes carrying the torch of generations before them. I began with these questions: What might climate solutions look like that simultaneously strive for a radical reduction in greenhouse gas emissions, adaptation to changing conditions, and justice for those most impacted by climate change and structural oppression? Amid a sea of siloed and shallow so-called climate solutions focused on scale, speed, and investor return, which solutions are most rooted in relationship, compassion, humility, trust, and long-term thinking? What does meaningful climate action and collective care look like from folks who have been here before, who have devised loving and innovative ways to meet the needs of their community members in times of crisis and shortage and oppression? This book emerged from those conversations.

Why Climate Resilience

The latest Intergovernmental Panel on Climate Change (IPCC) report confirmed that the climate crisis will continue to worsen until at least 2050, even if radical action to slash greenhouse gas emissions is taken today.[2] Unprecedented storms, floods, and heat waves are expected to become more frequent and intense, coral reefs are projected to face almost complete die-off, sea levels will continue rising, arable land and fresh water will become increasingly scarce, and an estimated one billion people around the world will be displaced.[3] Halting the expansion of fossil fuel projects and drastically reducing greenhouse gas emissions, particularly among the world's wealthiest, is absolutely imperative. And as humanity confronts the possibility of extinction, it is becoming increasingly evident that we must also learn how to keep one another safe as extreme weather becomes harsher and more prevalent and as basic necessities like food and drinking water become more unpredictable.

It is also becoming increasingly obvious that efforts to decarbonize will continuously fall short if we collectively fail to address the root of rising greenhouse gas emissions and the constellation of interrelated ecological and social crises that compound their impacts. Tethered to a growth-at-all-costs economic system of capitalism that forever demands *more, more, more*, all the renewable energy infrastructure currently being built only covers a fraction of new energy demand.[4] Corporations, which are legally beholden to prioritize their shareholders' bottom line, focus far more on marketing their peripheral environmental and social impact programs than on making meaningful changes. Substantive climate legislation continues to be undermined by politicians who are paid by the fossil fuel industry. And most global resolutions fail to spur action from the nations who emit the most greenhouse gases and refuse to potentially threaten their own global power and dominance—the same nations responsible for settler colonization and imperialism. Therefore, successfully evading climate catastrophe means truly confronting the cis-hetero-patriarchal paradigms of conquest, white supremacy, and endless accumulation underpinning this apocalyptic era.

Fortunately, countless women, two-spirit, nonbinary, and genderqueer folks around the country are already inciting a revolutionary shift in their communities, tapping into ancestral wisdom and centuries-old lifeways to resist the systems driving the climate crisis and to advance comprehensive climate action strategies grounded in compassion and care. They are demonstrating that, if we work thoughtfully to center liberation and healing in our climate solutions, and we allow those most impacted to lead the way, not only will our planet begin to cool again, but we also will all emerge that much more whole, safe, and free.

What You Can Expect from This Book

Climate Resilience is a collection of thirty-nine short essays edited from my interviews with activists, organizers, facilitators, ecologists, planners, earth workers, educators, entrepreneurs, strategists, farmers, healers, researchers, community leaders, artists, and storytellers who generously shared their work and wisdom. Collectively, their stories impart a comprehensive understanding of the root causes of the climate crisis as well as how it manifests differently across communities as climate impacts are compounded by chronic stressors like systemic racism, ableism, poverty, mass incarceration, food apartheid, pollution, and more. Their perspectives illuminate a path through the climate crisis and toward liberation, collective autonomy, and abundance for all.

The majority of the book's contributors are Black, Indigenous, and people of color, many are queer, and none are cisgender men.[5] This makeup is meant to reflect the grassroots climate movement as I have experienced it, as well as to center the voices of those who have been most impacted by the climate crisis and intertwined crises. I also hope that this project nurtures the feminist climate renaissance that the brilliant Dr. Ayana Elizabeth Johnson and Dr. Katharine Wilkinson helped put on the map with their groundbreaking text *All We Can Save*. White men, along with climate solutions rooted in more masculine ideals such as efficiency, competition, ego, scale, and domination, have been centered in climate conversations for too long. And these so-called solutions haven't worked. I'm eager to help amplify and usher in an era of climate leadership and solutions that are grounded in empathy, embodiment, generosity, interdependence, resourcefulness, gratitude, connection, and ease.

All of the contributors to this book reside in the so-called United States or on the land of their respective tribal nations on **Turtle Island**. (Please note that bolded terms throughout this book are further defined and contextualized in a glossary at the end of this book for deeper understanding.) The contributor makeup is certainly *not* reflective of the climate movement at large, as people around the world, and particularly those living closest to the land, have been sounding the alarm about climate breakdown and offering solutions for decades, while residents of the United States have been buffered from many of the most extreme climate impacts thus far. However, I chose to place a geographic limitation on the perspectives shared in this book to narrow its scope and respect my own constraints, as I only speak English fluently, my global climate network and understanding of cross-cultural nuances is insufficient, travel during a pandemic would be challenging, and so on. As a result, *Climate Resilience* lacks important climate stories and strategies from around the world.

There are so many incredible books and pieces of art and education created by global climate activists, and I highly encourage you to engage with and support their work!

These interviews have been edited for brevity and clarity. To keep this book as accessible and concise as possible, there is an abundance of brilliance that didn't make it into these pages. You can find more of the contributors' words along with additional interviews and resources on this project's companion website, www.climateresilienceproject.org.

Many of the essays are accompanied by deeper dives into specific climate strategies that the contributors are lifting up in their own work. Within these sidebars, I explain what these climate solutions can look like in practice and why they are important ingredients within an effective and intersectional climate revolution. I also briefly contextualize the solutions in their rich histories and offer some important caveats and general best practices around how they should be approached. Finally, I present suggested research and journal prompts for continued exploration and a range of action items that you can utilize in your own communities, designed to meet you where you're at, whether you have some spare time, space, or financial resources, or you have a particular skill set, passion, or community network to engage.

These sidebars are meant to introduce a multitude of potential points of entry into the climate movement. They are meant to inform, inspire, and ignite action! By no means are you meant to take on everything suggested in this book. The climate revolution calls for a true diversity of tactics, and not every tactic will speak to you. Please allow yourself to more deeply explore the solutions that resonate or make your heart skip a beat, and concentrate on the action items that feel like they will enliven rather than deplete you.

You'll notice that most action items focus on what you can do to make a meaningful difference at the local and regional scale. I recognize that individual behavior changes, like driving less, swapping light bulbs, or voting, can feel futile in the face of such massive challenges. I also understand that creating structural changes, like passing federal legislation or enacting global resolutions, is out of reach for most of us. I hope to challenge the disempowering myth that we must choose between individual or structural change and instead demonstrate how much is possible when people work together, collaborate thoughtfully and strategically, and optimize for the ripple effect.

You also may notice that many of the solutions highlighted have been around for centuries, if not millennia. The reality is that many communities have been cultivating resilience amid apocalyptic conditions for centuries. While humans are experiencing climate change on a global scale for the first time, white-supremacist settler colonialism, imperialism, and militarism have transformed countless landscapes and local weather systems around the world already. Myriad communities have endured the pollution, extraction, and enclosure of their land, water, and air, and millions of people have been forcibly displaced from their homelands or endured other forms of major community disruption. These stories can provide tremendous insight on how to resist and survive amid a version of apocalypse, and how to heal, rebuild, and find balance anew. It's a critical moment in which to honor and receive these stories and blueprints for action with immense gratitude and humility and to allow them to guide us through the coming chapter.

There is notably an absence of technocentric or market-based solutions in this book. Technology will be a necessary ingredient in our transition beyond fossil fuels. But it is too often touted as *the* solution, rather than a singular tool within a much larger societal reorientation. Plentiful other books explore the role of technology in climate mitigation and adaptation strategies more deeply. I hope that this text can be a respite from technocentric climate manifestos and false solutions that fail to adequately address the oppressive paradigms at the root of the climate crisis and related crises.

I also want to emphasize that this is not a complete guide to climate action. Climate solutions are infinite, and *Climate Resilience* highlights just a handful. The climate movement is dynamic and ever evolving, and by the time you read this text, it may be dated. This book is a small offering to expand our collective consciousness about what climate action can look like and where we might begin. It's a jumping-off point. I hope you're able to take what serves you, challenges previously held beliefs, sparks your imagination, and drives you to go *do*.

Who I Am

My fight for a livable planet began in 2005, when my seventh-grade science teacher introduced our class to the phenomenon of global warming. I was shocked that

I hadn't heard more about this seemingly very urgent and serious phenomenon, so I promptly created some poorly illustrated pamphlets to inform my neighbors. In high school, my climate activism manifested in the form of weekly greenhouse gas reduction tips sent out to my high school listserv (I belatedly apologize to my classmates and teachers who endured four years of this), building the school's first cafeteria garden with the environmental action club that I loved so dearly, and penning my first attempt at a book, *How to Be a Green Being*, a comprehensive guide to environmentally minded individual behavior changes that very fortunately never saw the light of day.

I came of age during an era when the prevailing ethos was that if enough people knew climate change was happening, if enough people believed that humans were causing it, and if enough people understood how they could shift their behavior to reduce their own carbon footprint (a concept popularized by fossil fuel company BP in 2005 in order to shift blame and accountability from oil giants to individuals[6]), we would halt the climate crisis. Over the years, and largely thanks to the labor and education of Black, Indigenous, and women and gender-expansive people of color, I have come to understand that evading climate catastrophe calls for revolution. It means changing everything. For a while, I felt frozen by the grandness of that sentiment, unsure of where and how to begin. Today, I see tremendous beauty, possibility, and hope in that sentiment and know it means that there are infinite ways to meaningfully engage in climate action.

In the past decade, my journey trying to find my place in the climate movement has been winding. I blessedly stumbled into the geography department in college, where my worldview was constantly challenged and expanded. I spent the better part of four years studying how well-intentioned (and sometimes blatantly self-interested and conniving) environmental, climate, and social justice programs can do more harm than good (particularly when involving wealthy and white-majority nations, organizations, and individuals intervening in communities other than their own). I graduated with more clarity around which so-called solutions didn't work, but I felt unsure of which climate strategies I should try to throw myself behind, for fear of inadvertently exacerbating issues or abetting in the delay of meaningful change.

I spent the next several years working on farms and in small businesses, at think tanks and nonprofits. I explored how to cultivate food systems rooted in justice, sovereignty, abundance, and beauty. I designed and built a tiny home and briefly studied green architecture before concluding that I was very bad at it. I went back to school to study intersectoral approaches to climate action more deeply, hoping that *the* climate solutions would reveal themselves, and I once again emerged with more questions than answers. I worked as a corporate climate strategist through graduate school and felt disillusioned and frustrated that most industries seemed to be shouting to the world that they were making real progress while tinkering at the margins and kicking the can down the road. (This is not to say that there aren't truly incredible people and awesome work happening in this space—it's just really hard to make meaningful change when the larger system isn't budging.)

For the past couple of years, I have been running a small, climate justice–focused foundation, engaging in climate education work, continuing to learn as much as possible, and spending most of my time trying to bring this book to life. Largely thanks to the incredible folks I've had the immense privilege of being in conversation with while working on this project, I now believe in an entire tapestry of climate solutions with my whole heart, and I'm so excited to share some of them with you. This is the book that I wanted and needed as a misguided but fired-up teenage activist, as a confused but eager college student, as a young graduate attempting to build a meaningful career, as an educator trying to construct intersectional and holistic curricula, and as a relative who has felt frantic to do more to protect her beloveds in the years to come.

I have often questioned whether I am the right person to author this book. I am keenly aware of my positionality and limitations as a white, cisgender woman with citizenship and class privilege. My gaps in understanding are numerous, and I am certain I will make mistakes within (and beyond) these pages. I thank you in advance for your patience in moments when I stumble. I want to emphasize that I do not take this responsibility lightly. I am endlessly grateful for and humbled by the contributors of this book, who entrusted me with their words, and I have worked hard to honor their wisdom the best I can. I know you will cherish their stories like I do!

Relationship Repair

Effective climate resilience strategies call for a radical restructuring of the ways in which we make sense of and take care of our collective home and one another. These essays are from activists, organizers, and facilitators who are challenging relationship models based upon extraction, exploitation, and disconnection—from the earth and from one another—and who are offering up relationship models rooted in reciprocity, reverence, and interconnection. They present a path forward, grounded in ancestral wisdom and millennia-old lifeways, to rehabilitate our relationships to the earth and our kin.

REVEREND MARIAMA WHITE-HAMMOND

SHE/HER/HERS

"Climate work is first and foremost spiritual work"

Spirituality often gets overlooked in climate conversations, but Reverend Mariama White-Hammond reminds us that it's foundational. Mariama is an advocate, faith leader, facilitator, and farmer. In April 2021, shortly after we spoke, Mayor Jim Janey appointed Mariama to the role of chief of environment, energy, and open spaces for Boston, Massachusetts, a position that will allow her to help accelerate efforts to enhance food security, create good green jobs for the next generation, find innovative ways to keep communities cool amid extreme heat, and move the city toward its goal of being carbon neutral by 2050. It's hard to imagine somebody more perfectly suited for this role, which was created with the expansive mission of enhancing environmental justice and quality of life in Boston while preserving and improving its architectural and historic assets. Mariama, a lifelong resident of Boston and champion of its people, grew up in a predominantly Black community in Roxbury and now lives two miles away in the third most diverse zip code in the United States; she intimately understands and knows how to solve for the social and environmental justice issues pertinent to the city's most impacted residents. She's also adept at connecting seemingly disparate issues and bridging communities to build cross-movement solidarity and advance true progress.

In fact, as the founding pastor of New Roots AME (African Methodist Episcopal) Church in Dorchester, Mariama shepherds a multiracial, multi-class community and regularly weaves issues of ecology, immigration, climate change, energy policy, and economic justice into her sermons. Social justice is central to the AME tradition, and Mariama follows in the footsteps of her parents, who are both community leaders and ordained ministers. Prior to going back to school for her master of divinity and seeding New Roots AME Church, Mariama served as the executive director of Project HIP-HOP (Highways Into the Past–History, Organizing, and Power) for thirteen years, joyfully engaging young people in activism, community organizing, and culture making through art.

Content Notes: Structural and institutional racism, mass incarceration, environmental injustice

On her journey into intersectional climate activism

I've always been an activist. I don't remember a time when that wasn't the case. I grew up around people who practiced **liberation theology**, and from an early age, I went to a school that was extremely focused on social justice. I remember being very young and deciding that I wanted to start a homeless shelter out of my house. My mom totally humored me, probably because she knew no homeless people were going to see a sign handwritten in crayons and think, *Great, I want to go there for shelter*.

When I learned about the hole in the ozone layer, I began advocating for people to give up their hairspray, which was a very big deal because it was the 1990s and many of those hairdos were only possible with tons of hairspray. And then I attended my first environmental meeting at my school. At that point, I was attending a predominantly white school that was two miles from my home but felt like light-years away. Everyone at the meeting was white and they were talking about polar bears and dolphins. I remember thinking to myself, *There are people dying*. The '90s were a really rough time in Boston in terms of violence, and these conversations about polar bears felt so disconnected from my context. Environmental issues were being framed from the mentality of conservation, and the history of conservation in this country is deeply tied to white supremacy.[1] The magnitude of the hole in the ozone layer had really spoken to me in the same way that I recognized the impacts of mass incarceration and the drug trade, but the environmental action space was so white. Eventually I thought, *These white people don't have to worry about their people getting locked up, they don't have to worry about educational equity, and they don't have to worry about all these other things, so let them deal with this one. I can let this go.*

The shift for me happened with Hurricane Katrina. I was running an organization called Project HIP-HOP at the time, and we had developed a lot of longstanding relationships with communities in New Orleans. I felt connected to a lot of these neighborhoods, and when I saw what was happening, it was heartbreaking. One particular story that really stuck with me was about a jail that didn't have an evacuation plan. The guards were getting ready to go home to get their families out, and there was a point at which they were just going to leave the prisoners in their cells. They would've died. They eventually let them out, but it really struck me because my cousin was in prison at the time and so many of the young people we were working with were caught up in the criminal justice system. I kept thinking about what would happen if the guards had just walked off and someone like my cousin was allowed to drown in his cell. At that moment, I knew that I couldn't let well-meaning white people be the only ones working on climate change. I knew they wouldn't be thinking about evacuation plans in jails. They wouldn't be committed to radically transforming our economy in the major way that was needed while supporting the people who have been screwed by the economy as it currently stands. These kinds of issues weren't going to be addressed by the people I saw actively working on climate change at the time.

When Kimberlé Crenshaw coined the term ***intersectionality***, she was really talking about the reality of Black women's lives. For me, intersectionality isn't a philosophy or something that I've picked up along the way. It's navigating between communities as a Black woman. I consider myself a fierce womanist, but there are challenges working in that community. I consider myself an LGBTQ+ ally, and that gets me into some challenges in religious communities. And on and on. Climate organizing has created opportunities for me to build relationships with other communities, but that's always been my inclination. I feel deeply nurtured by many communities, some of which do not relate amazingly well together, and I feel like it's part of my work to help people see that there are both lots of connections and lots of challenges.

On climate action as a spiritual imperative

When I was younger, I used to wish that I was born during the civil rights movement. But I started to realize that this is our moment. My spirit started saying this is an existential crisis for humanity. We'll either rise to the occasion or not. I try not to spend too much time thinking about the *or not*. I know it's a higher percentage chance that we end up in the *or not*, but what does it benefit me to think about that? I have to put all my energy behind the possibility that we as humans will be who God, I think, always wanted us to be. And if not, at least I will have done my absolute best to get us there.

I'm overwhelmed by the immensity of the challenge, but I've also come to understand that if we have to make a radical shift, we also have the opportunity to rebuild the world better than it is, not just to try to survive a catastrophe, but to actively create something worth laying it all on the line for. For me, it's not worth laying it all on the line just to get to 350 parts per million of global atmospheric carbon dioxide.[2] This moment is about a fundamental reorientation of our relationships.

When people ask me what I do, I say that I do ecological justice work, rather than climate justice or environmental justice work, because *eco* means home and we've got dysfunction in our home. We need to shift how we're relating to other humans and how we relate to all other beings on whom our lives depend. Right now our relationships are abusive and destructive. Quite frankly, when you look at humanity from the perspective of other living beings on this planet, we don't really deserve to be here. We're a hot mess. We don't know how to act. We're the worst houseguests ever. Our work is to bring justice to the relationships in this home. And as far as I'm concerned, despite what folks like Elon Musk may argue, we only have one home.

Therefore, climate work is first and foremost spiritual work. Technocratic fixes can't address the way that we relate to each other, to the earth, and to other species on this planet. Our relationships are the source of our problem. I'm not anti-technology, but I'm frustrated by people who believe that technology will save us. *We* have to change. Technology may assist us in our reorientation, but if we believe that we can just find technologies that allow us to live the same but with a lower carbon footprint, that, in my opinion, is a pipe dream.

We're staring down the biggest evolutionary shift for humanity, ever. I do believe it's within us to be better. I do believe that we want to be better in the depths of our hearts. And I know that, in moments of history, enough of us have become our best selves to allow major shifts to happen.

RUTH MILLER

SHE/HER/HERS

"The fight for climate justice and Indigenous rights is indistinguishable from my own healing"

Ruth Miller was also bestowed the name Łchav'aya K'isen, which aptly means "whirlwind woman" in Dena'ina. While speaking with her, I could not stop thinking, Ruth is a force, both in the ways that she is able to deftly speak about climate injustices, Indigenous sovereignty, youth activism, and future possibilities with ease, and also in the ways that she rigorously commits to living her values each day. She is a Dena'ina Athabaskan and Ashkenazi Russian Jewish woman and member of the Curyung tribe. In 2019 Ruth graduated from Brown University, where she studied Indigenous resistance and liberation, led the student group Natives at Brown, and was the university's first Native American student to speak at the commencement ceremony. Since then, Ruth has returned to Anchorage, Alaska, attended several international climate conferences, and blossomed into the climate justice director for Native Movement, a matriarchal, grassroots, Indigenous-led organization that advocates for Indigenous peoples, lands, and waters and fights for justice for future descendants and ancestors. Whether advocating for Indigenous sovereignty and climate justice at the local, national, or international level, Ruth is incredibly thoughtful about how she carries her peoples' voices with her and how she can transcend stale conversations about climate jobs and market-based solutions to create expansiveness for **Just Transition** solutions grounded in reciprocity and care.

Ruth is also a singer, storyteller, traditional beadwork artist, and subsistence fisherwoman. She is a steward of the Dena'ina language and is on the path of becoming a traditional healer. In all that she does, Ruth centers Indigenous values and lifeways and carries with her radical love, softness, and a deep commitment to kin.

Content Notes: Settler colonization, Indigenous genocide, Boarding Schools, sexual assault, and gendered violence

On her own journey into climate justice organizing

I was blessed and lucky enough to grow up in a family of loud, well-spoken activists. Both my parents are Native American rights lawyers, but they each approach this work from extraordinarily different circumstances.

My mother is an Indigenous woman who was a high school dropout and became a young mother to my sister at twenty-one years old. Eventually, after working as a laborer on the oil pipeline, she decided to go back to school. Once she graduated valedictorian, she was awarded a full ride to Harvard Law School. She got herself through school with her daughter by her side and became one of the forefront Native voices in federal law history.

My father is a Jewish man from New York City who grew up with tremendous access and exposure. Education came easily to him. However, he turned his passions, attention, and tireless work ethic toward public service for Indigenous rights. Through a very different approach, he began his long legacy of advocacy for tribal nations, securing some of the largest financial reparations from the US government for Indian Country.

While I was raised with these two incredible role models for what it means to defend basic human rights, my family also carried a lot of generational trauma induced by colonization and perpetuated by the climate crisis. On the one hand, I grew up fluent in and speaking often of the fights we were waging and the causes we were advocating for. On the other hand, I lived through the very emotional and raw experience of substance abuse, alcoholism, and trauma present in our own family and our own needs for healing. It made the fight for climate justice and Indigenous rights indistinguishable from my path and my own healing.

I can't imagine not doing this work, not for lack of interest in many other things, but because we are confronting a global ultimatum. To me, climate justice isn't my job. It's a nonnegotiable responsibility to my own community and to my descendants, my future children and grandchildren. Every person should be devoting their utmost capacity toward addressing climate justice. Fundamentally, it is not only the determinant of humanity's collective future, but it also defines the kind of world we want to manifest and the kind of world that we're willing to live in. But if everyone cared about our planet and our collective liberation enough to fight for equality, not just across race and class but also of all life on this earth, including our lands and waters and nonhuman relatives, I guess we wouldn't be in this mess in the first place.

As a community organizer and as a young person, I must admit that I am existentially exhausted from the circumstances of our world. I say that as someone who has a job, a home, financial security, and food on my table. And I'm tired. I think that I have a very deep-seated impatience to manifest this liberated, beautiful, loving vision that we're all fighting for every day, because I'll admit, it seems far away. I must remind myself that I'm twenty-three years old, and I might not live to see the world that I want to create. That's a very present reality. I have to ask myself, *How can this process be successful without ever reaching ultimate objectives?* That doesn't mean I'll fight any less for it. But that does mean, if I'm going to dedicate the rest of my work to this fight, that I must find space for joy, rest, and ease. I must understand how important it is to practice radical compassion with myself and how liberatory it is to laugh in community. And I must see the success in the process.

As I strive to teach myself patience, moderation, and forgiveness, I also have the joy of reflecting on the fact that this is hard because the things that we're fighting against are meant to make our lives hard. Of course it's hard.

It's frustrating that the burden of this advocacy so often falls on those most affected by it. Those who are fighting the hardest are also the ones who are targeted and suffering the most. There's a spiritual exhaustion that comes from fighting things that are impacting you on a deep and personal level, that you can't just turn off when you close your laptop. The world can be a very tiring place for those deep in these fights for justice.

I find nourishment to keep going when I spend time with community. When we're able to stand together, sing, drum, share food, laugh, and talk story, and have our elders present and have kids running around, when we are able to battle the darkness that we're confronted with using our community spirit, our collectivity, and our strength to ground us. That is the purest form of power.

On the climate crisis and intertwined crises in Alaska

In really broad strokes, when I think of the impacts of climate change here in Alaska, I think first and foremost of our frontline and rural communities that don't have the privilege of escaping the day-to-day impacts of the climate crisis. Due to the phenomena of **Arctic amplification**, communities in the Arctic and circumpolar North are experiencing the climate crisis at a disproportionately faster rate than the rest of the world.

They're experiencing thawing **permafrost**, which is destabilizing our lands and infrastructure, and melting sea ice, which is forming weaker and thinner than ever before and is no longer able to protect our coastline communities from winter storms. They're watching dead salmon wash up on our shores and our riverbanks, if they even make it up the river to their spawning waters at all. They're experiencing erosion that is making homes fall out from underneath their feet. The impacts of climate change are affecting every facet of life, whether that be food systems, housing and infrastructure, or physical health and well-being. And so our Indigenous communities, who are often left without support even in the best of times, who have lived closest to the land for millennia, have become some of our world's first environmental refugees.

When I consider how the climate crisis has impacted my community, I also reflect upon my personal experience of the ways that Alaska has been used as an oil colony to the US empire.[1] Not only is the history of Alaska one of illegal occupation, falsehoods, and unestablished claims to land, but the worldviews, methodologies, and systems that accompanied settler colonialism in Alaska are fundamentally at odds with the Indigenous ways of life that still persevere here.

Our climate crisis began when settlers came to our lands and saw it as something to be used, taken, and exploited, instead of something to be honored. The exploitation of our lands, waters, resources, and animals is the very foundation of colonialism and, with it, the global capitalism that demands a fossil fuel–based economy.

In spite of the growing crisis, increasingly risky and desperate development projects continue to disrupt our ecosystems. We've seen a huge increase in respiratory illnesses and cancer in our Native populations within the proximity of development projects. We see direct impacts on our food systems. For instance, the opening of the Arctic National Wildlife Refuge to increased oil drilling will have potentially devastating impacts on the Porcupine caribou herd, the heart and lifeline of the Gwich'in and Iñupiat peoples of the Western Arctic. We also see direct devastation to our Indigenous women. To support mining and oil projects, "**man camps**" are built to house imported labor—men shipped in from outside Alaska. Near such camps, we observe a proven and consistent rise in violence against women and Two-Spirit relatives. Alaska has the highest rate of domestic violence and homicide in the entire country. For Indigenous peoples, the issue of defending our bodies and our lives is intricately tied to ending the industries that contribute to the current climate crisis.

So when we think of climate justice here, we're not just thinking about ameliorating the very real lived impacts of our frontline communities that are left the most vulnerable to the climate change that they had no part in causing. We also have to begin to question the systemic global frameworks that are predicated upon the sacrifice of our Indigenous lives and homelands.

On the persistence of Indigenous sovereignty

To understand modern Indigenous sovereignty, we have to understand that all settler colonization on this continent was founded on the erasure and genocide of Indigenous peoples, and only in very recent decades. In Alaska this erasure and genocide is within our living memory. Our grandparents and parents still hold the trauma of being taken away from their families and sent to boarding schools where they were whipped, beaten, and violated for practicing their own cultures and for speaking their own languages.[2] When we begin to question the dominant American education system, which tells us that Indians are dead, that Native culture doesn't exist, that we're all just citizens of the United States, then we begin to understand the very shaky and fragile foundations that undergird the United States empire.

When we question the process of exploitation and oppression, it reveals to us that there was nothing inevitable about the domination of the United States. In fact, we have other alternatives that have been preserved through our community resilience and a commitment to real justice. We have the shared memory of our trauma, but we also have the shared memory of tens of thousands of years of successful economy, successful society, successful community health and public health, successful education systems, successful community policing systems, and successful Indigenous science that provided a model for a symbiotic and mutually beneficial relationship between people in the lands. During those tens of thousands of years, we did not encounter a global climate crisis at this scale. We did not encounter systemic oppression at this scale because our methodologies and worldviews came from the land and our interconnectivity as beings of integrity. Through our creation stories and our spirituality, we were taught the benefits of living in reciprocity with lands and waters. That isn't an aspect of modern colonial United States and global capitalism.

When we talk about Indigenous sovereignty, what we mean to say is that Indigenous peoples are sovereign peoples, and tribal nations are sovereign governments, both ancestrally and legally under the laws of the United States. To achieve Indigenous justice, we need to begin to interrogate the ways that our sovereignty is being denied. Indigenous sovereignty is being denied through our limited access to our own resources: our own hunting, fishing, and subsistence rights. It's being denied through our underfunded schools and healthcare clinics and through voter suppression. It's being denied through the over-policing of Black and Brown communities, and particularly in the territories of tribal nations.

To navigate Indigenous justice, we must confront the day-to-day ways that our communities are continually oppressed and kept vulnerable and susceptible to violence and poverty. Additionally, we have to think about achieving justice on a spiritual, methodological, and epistemic level. We deserve sovereignty of our minds and of our spirituality. We deserve to live in a world that is not built upon the rape and violation of our lands through resource extraction that our people have not consented to. When we begin to remind ourselves that history is much longer than the white colonial history that we've been taught, we see that there is a much longer, stronger case for symbiotic, sustainable relationships with lands, waters, and resources. We do not consent to the way that this world currently profits: for the very few, at the expense of the very many. We know and remember a time when this was not so, and we demand a return to it. That is what justice means.

NIRIA ALICIA

SHE/THEY/WE

"My responsibility is to my ancestors and their seeds"

Niria has created profound ripple impacts in the climate justice movement. She is a United Nations Environment Programme (UNEP) Young Champion of the Earth, recipient of the GreenLatino Emerging Leader Award and North American Association for Environmental Education 30 under 30, and a former COP25 delegation leader with SustainUS, a youth-led climate justice organization. She is a Xicana human rights advocate, educator, storyteller, and a core organizer of the Run4Salmon prayer, a spiritual journey to restore the Chinook salmon runs and Indigenous lifeways. She is also an organizer and strategist for numerous other climate justice campaigns by groups like Honor the Earth, Women's Earth Alliance, Greenaction for Health, and Earthjustice. There is no doubt that the climate movement is deeply enriched by Niria's presence.

Perhaps more impressive than any singular accomplishment or accolade, though, is how seriously Niria takes her responsibility as a relative—to her family, her ancestors, the many communities that she is a part of, and more-than-human worlds. She leads with reverence, ritual, care, and radical love. At a pinnacle moment in the Stop Line 3 movement, as thousands of people from around the world gathered at the headwaters of the Mississippi River to stand in solidarity with water protectors and protest the construction of the Line 3 tar sands pipeline through Anishinaabe land and waters, I had the honor of witnessing Niria Alicia lead the masses in song and prayer. Her words became a crucial source of unity and strength for protestors persisting through throngs of police and thick summer heat.

Content Notes: Settler colonization, species extinction, domestic abuse

On her lineage as a land and seed steward

I was born in Medford, Oregon on Takelma land in 1992, 500 years after Columbus invaded Turtle Island. I believe my ancestors sent me here that day to turn the tide for the descendants who will come after me 500 years from now.

By way of my mother, my roots are in Michoacán. She learned how to read the moon and work the land from my great-grandfather Maximiliano. He taught her how to farm and live with the seasons. She learned how to make fire and cook and prepare our ancestral foods and medicines from my great-grandmother Leobijilda. She came to the so-called United States to work as a farmworker and restaurant worker and to this day continues that work and maintains our family's ancestral seeds and recipes.

By way of my father, my roots are in Chihuahua. In his days as a tree planter, he came to plant millions of trees in the states of Washington, Oregon, Idaho, and California. My parents came together on Takelma land, in an area now known as the Rogue Valley. I am their oldest daughter. When you come from a culture where women run the families, then you know that's a big responsibility. My mother is the oldest daughter and my grandmother Alicia was, too.

I grew up in a migrant farmworker community, and I was a migrant farmworker child. I would harvest grapes on the weekend and clean houses on short school days. When I got old enough to legally work, I packed peaches.

I am also an heirloom seed keeper. My family on both my mother's and father's sides have planted and stewarded corn, pumpkin, bean, and chile seeds for generations. When my great-aunt came to Oregon, she brought some of our family seeds with her. Naturally, the corn from Michoacán didn't immediately acclimate to southern Oregon's soil and climate. She'd plant a whole field and only a couple of seeds would grow. But she'd take those seeds and would plant them the next year. She'd do the same thing again and again, until after many years she was able to climatize our seeds to the soils of Tacoma land. It's my responsibility to keep and steward those seeds.

Coming from a farmworker community, and from a seedkeeping lineage, where you're so close to the land, the elements, and the plants, it's hard not to become drastically aware of the shift in animal and weather patterns. My understanding of climate justice traces back to when I was thirteen years old, waiting for phone calls to hear if temperatures would drop and we would have to go out into the orchards to light the heaters so that the pear blossoms wouldn't freeze.

We know that our communities contribute least to climate change but are paying the highest price. The first people who are impacted by the climate crisis and who are most deeply impacted are the people who are connected to the land, like our Indigenous people, who are always on the front lines against extractive industry, protecting our land, water, and air, and like our farmworkers, who are out there working the land to feed the people every single day. We understand that climate change is a manifestation of capitalism, colonialism, and patriarchy, and the extraction of Mother Earth that occurs when society is so disconnected from valuing and honoring the sacredness of Mother Earth. At the same time, we recognize, and the United Nations recognizes, that most of the world's remaining biodiversity is in the stewardship of Indigenous people. I have to fight for that. If I'm going to make my life worthwhile, I have to fight for Indigenous rights, for the rights of my community as migrants, and for the Indigenous people on whose land I live.

On sacred interconnectedness with salmon

My great-grandmother told me that I have a responsibility to protect the wellbeing of the waters on whatever lands I call home. From Alaska all the way down to Mexico, salmon are caretakers of the waterways. Our salmon relatives are genetically, spiritually, and culturally embedded within the land. Their presence is extremely important, and so their absence is really alarming. This keystone species is on the verge of extinction. The Winnemem Wintu have a prophecy that when the salmon are gone, the Winnemem Wintu people will also be gone. *Winnemem Wintu* translates to "middle river people." I wonder if the Winnemem elders were referring to all of humankind.

Salmon have been around for millions of years. They're so wise. As humans, we need to humble ourselves to our relatives that really know what it takes to stick around Mother Earth. We're not doing a good job. We have been around for a fraction of that time and we've caused so much harm.

If we want to build a future where we can stand a chance to survive the catastrophes on the horizon, it's really important that we build back with salmon and with these elder species. Chief Caleen Sisk, the tribal chief of the

Winnemem Wintu Tribe, says that we have to follow the salmon. And what do the salmon do? They do things they've never done before and they follow the stars. We have to do things we've never done before because we've never before been here, on the brink of extinction. We need to learn how to leave the creeks of our comforts and go out into the ocean, knowing that's what our species needs to survive. When the salmon leave the comforts of their tiny creeks to go out into the massive ocean, they don't know if they're going to come back. But they follow the stars and the Creator. Some of those who make it will return home with a stronger Spirit and more wisdom, and those are the ones that spawn. We have to strengthen ourselves in that same way. If we follow the salmon, they'll show us what to do.

Ultimately my responsibility is to my ancestors and our seeds. In order to grow and thrive, those seeds rely upon the wellness of water and the territories that I call home. The wellness in those territories relies upon the presence and wellness of the salmon relatives. So that's how my struggle for liberation is interwoven with the struggle for the protection of water and the restoration of salmon for future generations. Life begins and ends with water.

RESILIENCE TOOL SPOTLIGHT

Seedkeeping

For more than ten thousand years, it was customary for people to harvest and save seeds as they grew food each season. This practice allowed seed keepers to preserve and pass down incredible genetic biodiversity, ancestral wisdom, and cultural identity, and to ensure that their communities would eat well for generations to come. They optimized for taste, texture, fruit size, hardiness, productivity, color, shape, disease and insect resistance, seasonality, and more, adapting their seeds to the local growing conditions over time and growing ever-tastier varieties. There are also countless examples of seed saving as an act of resistance, sovereignty, and survival—of West African women braiding the seeds of sugarcane, yams, collard greens, and rice into their hair before being violently expelled from their homelands, and of Cherokee peoples carrying deep purple and jet black pole beans in their pockets along the Trail of Tears as they were forcibly displaced from their ancestral lands.

Over the span of the twentieth century, as millions of relatively small, diversified family farms and bountiful gardens gave way to fewer, larger, mechanized, monoculture farms, many food growers stopped saving seeds and started buying commercial seeds. Industrial agriculture giants like Monsanto, Dow, and DuPont developed hybrid seeds (which cannot be resown) and patented their biotech seeds (which cannot legally be saved). Seed saving eventually became too legally risky even for farmers who were still growing food from their own seeds; if their heritage seeds were contaminated by the patented seeds from a nearby farm, they could face a costly lawsuit. In 1930 a mere 0.5 percent of farmers planted commercial seeds. Today that number has rocketed to over 90 percent.[1] During the same span of time, we've collectively lost more than 90 percent of edible plant varieties.[2] When genetic biodiversity is lost, it's lost for good.

Today seedkeeping is a powerful way to preserve genetic biodiversity, to reclaim and reconnect with ancestry and culture, to build independence from transnational agriculture corporations and interdependence with community, and to help local food systems adapt to a changing climate. With just a handful of seed varieties per crop, food supply is particularly vulnerable to extreme climate conditions, insect infestations, and crop diseases, all of which we can expect to be exacerbated by social, political, and environmental threats in the coming decades. For instance, the Great Potato Famine of the mid-1800s devastated Ireland, where the vast majority of potatoes were of a single variety susceptible to blight. Meanwhile, Quechuan communities in the Peruvian Andes, where the disease was said to originate, knew to grow hundreds of potato varieties, understanding that even if some varieties didn't do well, the community would never go hungry.[3] Plant biodiversity is an insurance policy that positions us to adapt and continue feeding our communities through crisis.

To engage with the seedkeeping movement:

- **Research prompts:** Which foods and medicinal plants were meaningful to your ancestors, and which of those plants might grow well where you currently reside? Which foods and medicinal plants are meaningful to the original stewards of the watershed in which you reside, and which people or organizations are working to return seed ancestors home to their peoples and reclaim seed sovereignty? Are there any seed saving libraries or seed exchange groups in your community that you might be able to tap into?
- **Journal prompt:** In growing a garden and saving seeds with intention and care, which qualities feel the most important to hone and preserve to nourish, heal, connect, and delight community for generations to come?
- **If you have access to space to grow plants:** Start a seedkeeping practice at home or at your local community garden. Begin by planting open-pollinating seeds (meaning that they produce offspring identical to the parent, unlike hybrids), which are oftentimes marked "OP" on their seed packet, and choose self-pollinating crops, like beans, peas, lettuces, and peppers, while you're getting comfortable with the process. Try to plant just one variety in a species to prevent cross-pollination and preserve pure seeds. When seeds are ripe, oftentimes after the crop has matured, select seeds from the healthiest plants and consider which qualities you'd like to optimize for (i.e., save seeds from the plants that didn't experience disease infestation, that produced the most fruit, that tasted the yummiest, etc.). Clean and dry seeds. Store dried seeds in paper envelopes or seed packets labeled with the variety name and date of harvest.
- **If you have access to time and a passion for seeds:** Visit, volunteer at, or start a seed library in your community. Seed libraries are set up

for patrons to check out seeds for free and return well-labeled seeds for others to check out. They help cultivate reciprocity and interdependence, local biodiversity, food system resilience, and seed accessibility for all. If you're helping organize or set up a seed library, consider sectioning seeds by level of difficulty to grow; ensure that seeds are well labeled; make instructions for use highly visible, ideally in multiple languages; and get the word out so that community members know the seed library exists.

- **If you have access to a robust community network:** Host a seed swap with community members to exchange seeds and introduce seed saving to folks new to the practice. Consider beginning the swap in a circle, expressing gratitude for the seeds and the ancestors who saved them, and sharing more information about the purpose and logistics of the event.
- **If you have access to financial resources:** Help fund the efforts of seed keepers and organizations who support the reclamation and preservation of seed sovereignty, advocate for the legal protection of heirloom seeds, and help facilitate seed saving and exchange.
- **General best practices:** Seeds are sacred. Seeds are ancestors. Treat them accordingly.

MORGAN CURTIS

SHE/HER/HERS

"My people have a particular accountability for what's happening right now"

As a descendant of first and early white settler colonizers of so-called New England, Morgan Curtis has felt guided by the call to transform and transmute the legacy of her ancestors by organizing folks with shared backgrounds to redistribute wealth, repair white ancestral harm, and heal relationships with community, land, ancestors, and self. She supports individuals and groups through this process as a facilitator, ritualist, and money coach. Morgan's approach is largely shaped by frameworks and practices that she has lived and been mentored in as a longtime resident of Canticle Farm, a multiracial, interfaith, cross-class, intergenerational, intentional community. She is also informed by her community at Harvard Divinity School, where she is focusing on the spiritual dimension of reparations work for descendants of colonizers and enslavers.

Morgan has been a gentle but mighty force in the climate justice movement since she was a teenager. In fact, Morgan was in the grade above me at Dartmouth College, and when I arrived on campus, she was already a legend in the climate community at school and in the larger Upper Valley. Among her many efforts as a student activist, she cofounded a fossil fuel divestment campaign that was ultimately successful after eight years of diligent organizing. After graduating, Morgan engaged in countless struggles for climate justice and sovereignty around Turtle Island and beyond. She was instrumental in organizing folks to defeat a proposed oil drilling project in the Isle of Wight in the United Kingdom, where she spent part of her childhood. She once spent five months biking 3,000 miles to gather grassroots climate action stories on her way to COP21 in Paris. For several years, Morgan organized with SustainUS, a youth-led climate justice organization, honing her understanding of the global struggle for climate justice as well as her place in the movement.

Content Notes: Settler colonization, slavery

On her journey into collective healing and reparations work

I was a little kid that was very upset about the state of our earth and humanity's relationship to it. I remember reading an article in *National Geographic Kids* magazine about the giant industrial project of the tar sands in Canada, and just being so distressed about the devastation that was happening what felt like a world away. I went into the kitchen clutching the magazine and I tearfully asked my father about it. He said to me, "Huh, that project is very profitable. Your grandfather actually invested in that." And that was the first of many moments for me of realizing that the many privileges and comforts I was born into were connected to the suffering of others and to the plight of our planet in a time of ecological devastation.

From being that little kid upset about climate change, I came to college and decided to be an engineering major. I guess I thought we were facing a technological crisis. That was the narrative I'd been told, that we're using one set of machines and we need to be using a different set of machines and then everything will be okay. Not that long into my engineering major, I saw that that was pretty clearly not the case. I thought, *Okay, if this is not a technological crisis, I think it might be a political crisis. I think it might be a cultural crisis. I think it might even be a spiritual crisis.*

What does it mean to engage with a political crisis, a cultural crisis, a spiritual crisis? It doesn't take technology. It takes engagement with who we are as individuals and as a collective. I had always been told that humanity is doing something wrong, but over time that narrative was fine-tuned. It's actually a particular *thread* of humanity. And in particular, my people. People in the United States, people with wealth, white people, people who are investors, people who are rich, people who are educated in elite institutions. These are the people who are not only responsible for the most greenhouse gas emissions, but have actually set up the economy to be dependent on extraction in the first place.

I was asked over and over again by my fellow climate organizers, youth of color, *Who are you? And what are you doing here? And who are your people? And can you please go organize them?* I came to realize that my people, as shown by my family history, have a very particular accountability for what's happening right now. And we're not going to be able to move forward into any kind of just, sustainable future until we've reckoned with how we got here, which was through a lot of violence against Indigenous people, against Black people, against other nations through imperialism around the world, and through violence against our earth, our living planet. We have to heal from that. It's urgent. A teacher of mine, Joanna Macy, says that in the time of the Great Unraveling, we have two choices; we can turn toward each other or we can turn on each other. We are not going to be able to turn toward each other without reparations and truth telling for what has come before us.[1] There is no collaboration between the people of the earth at this time without acknowledging what has happened between us. So that is where I find myself in my work at this moment.

On reparative action and wealth redistribution in practice

First of all, I'm still learning what personal reparations and wealth redistribution should look like. It's not something we can do alone. But it involves research, truth telling, stopping the harm, turning toward repair and return, and building the new.

#1 and 2: Research and truth telling. I think this process begins with being really willing to be quiet and listen and learn, to rigorously research and be willing to tell the truth about our families and what they've done and what has been done in our name. This means both uncovering whatever family history there is to uncover, and it means looking into our bank accounts and stock portfolios for what is being done on our behalf right now.

My research and truth telling process looked a lot like uncovering my family history of colonization, being descended from some of the first and earliest settlers of the Northeast, uncovering my family history of enslavement, debunking the myth that slavery was only in the South, uncovering where ownership of Indigenous lands contributed to wealth accumulation in my family, uncovering my great-great-grandfather's co-ownership of a sugar plantation in Cuba, uncovering my great-great-uncle's directorship of a gold mine in Honduras, on the leading edge of US imperialism there, and then uncovering the investments that had been made in my name for my college fund in the fossil fuel industry, in the weapons industry. I know that fund was set up with the understanding that it was what was best for my future, and I honor that impulse, but I'm choosing to redefine what's best for my future.

#3: Stopping the harm. Stopping the harm means divesting from those systems to the very best that we can. Knowing that we are individual participants in a system that is much bigger than us, we can still make choices to participate less in the extraction and domination inherent in almost all our economic choices right now.

Stopping the harm, for me, was divesting from the stock market, seeing that there was no good way to benefit from the extraction of labor and resources. I'm choosing to not participate in the system of private property here on this continent. I do my very best to live simply and lightly. My living expenses living in community are about $20,000 a year, and it has felt really true to me that the less I spend, the less I'm contributing to a culture of consumerism and accumulation.

An elder in our community, a Dakota man, was telling me over the weekend that every year, when he goes to Sun Dance, an annual ceremony in his community, he gives away all of his money and leaves the ceremony with zero. That's his practice every single year. And I think about how my worldview and my culture tells me to always accumulate, to always have more. And his practice of his culture tells him that the highest expression of their values is to give it all away. I'm not there yet.

#4: Turning toward repair and return. Turning toward repair and return has looked like redistributing the $600,000 I inherited, as well as half my income from the work I do supporting other people with wealth to take action towards redistribution. My focus has been on resourcing Black- and Indigenous-led organizing and land projects, as well as moving money through mutual aid networks and in my immediate community. I also give to national movement formations

and regranting organizations that give not only the money but also decision-making power to Black, Indigenous, and other communities of color about where resources should go.[2]

It has also meant looking at where I am personally accountable for specific acts of healing and repair. For instance, one of my ancestors, whose portrait hangs in the hallway of my childhood home, was a general in the Second Seminole War, which was an act of genocide against the Sminole and Maskoke people in the Southwest of this country. Actually, before I even learned that fact, I had already felt called to return resources to a particular group of Muskogee people returning to their homelands. So I think it's also about trusting the impulses and intuitions that call us, being willing to take risks and respond to what we're being asked to do, even when we don't know if it's exactly right, even if we don't know what will come of it.

For the harm that has been done by my ancestors in Honduras, I've been moving resources through a regranting organization called Thousand Currents that moves money to grassroots-led social movements in the **Global South** struggling for land defense and food sovereignty. I recognize that, when we're talking about climate change, we need to be following the leadership of folks in the Global South who are most impacted by the crisis. That was something I profoundly learned from my time as a youth climate organizer in the international process. And I recognize that as white, wealthy people in the **Global North**, we have basically no capacity to figure out a good way to do that ourselves, and so we need to move our resources to structures and entities that build long-term, accountable relationships of mutual support with movement formations in the Global South.

I also acknowledge that I'm participating in ongoing colonization as a settler on unceded Lisjan Ohlone land today. I have been able to both financially resource and organize for direct land return to the Ohlone people through the Sogorea Te' Land Trust, and that has been a big honor.

#5: Building the new. As we move resources in response to the harms of the past, we also must move them to build the system that will carry us through this transition. In some cases it might look like investments and lending and creating circulations of capital that can support self-determination for communities that have not had that under racialized capitalism.

I think we're really blessed to be alive in a time when so much incredible creativity and imagination is being put toward building a **solidarity economy** that's based on reciprocity and gift and meeting the needs of all. How do we create an economy that works for life? I look to the leadership of places where that reimagination work is happening, such as Movement Generation, the New Economy Coalition, and Seed Commons.

We see a lot in the mainstream media about climate solutions that are promoted by predominantly white, male entrepreneurs who have an idea, a solution, whether it's a machine or an app or something else. I love them for their noticing that change is needed. But I know that for those of us who were raised in a capitalist, white-supremacist culture, and who are beneficiaries of that culture, the ideas that we come up with are so dramatically hindered by our indoctrination in that system that it is not a smart investment to put our money behind those ideas. As people who were raised by an education system that tells us we're the best, we're the smartest, we have what it takes to change the world, a lot of what we need to do is unlearn that and recognize we were actually trained up in a pretty damaging way. We need to be willing to set down that idea, that the way that we will change the world is by creating some big idea and gathering resources for our big idea and imposing it on others.

On the words she regularly returns to

A mentor of mine, Anne Symens-Bucher, cofounder of Canticle Farm, often says, "Whenever I notice myself saying, *I can't do it*, I just add an *alone* to the end." So if you're feeling like, *I can't do it*, it's probably true. You can't do it *alone*, but you could do it with others. We all need to find our people, and get to work together.

CASEY CAMP-HORINEK

"We are not these big protectors of nature; we are nature protecting itself"

Casey Camp-Horinek, also called Zhuthi, is an internationally renowned environmental ambassador, Emmy award–winning actor, author, elder, and the hereditary Drumkeeper for the Ponca Pa-tha-ta Women's Society. However, Casey identifies first and foremost as a matriarch, survivor, mother, grandmother, great-grandmother, sister, daughter, auntie, wife, and future ancestor. Casey self-describes as a product of Generations of Living in Tune with All That Is, and also a product of the disruption in the late 1800s that led to the genocide and forced removal of ancestors. In all that Casey does, Casey works to serve and honor all relations, reclaim traditional Ponca values and practices, restore environmental balance, and resist extractive and oppressive structures. (Casey prefers not to name pronouns, as pronouns were absent from Ponca language prior to colonization.)

Casey's environmental justice advocacy began many decades ago while bringing attention to and seeking to remediate the corridor of toxic environmental industry surrounding the Ponca Nation of Oklahoma. Since then, Casey has spoken to the United Nations Permanent Forum on Indigenous Issues, marched on Washington, DC, with Moms Clean Air Force, protested the Dakota and Keystone XP pipelines, joined the boards of Movement Rights and Women's Earth and Climate Action Network International (WECAN), and lent support to countless grassroots environmental movements around the world. Casey has been particularly instrumental in stewarding the Rights of Nature Movement, an effort to restore harmonious relationship with nature via a framework and legal theory that acknowledges the inherent rights of ecosystems and species to exist and thrive. Casey was instrumental in the drafting and adoption of the

International Indigenous Women's Treaty protecting the Rights of Nature, and after being elected to the tribal council in 2016, Casey helped the Ponca Nation become one of the first tribes in the country to enact Rights of Nature law to help put a moratorium on fracking and injection wells. I had the pleasure of listening to Casey speak about Rights of Nature at a conference years ago, where I watched as the auditorium became abuzz with awe and possibility. It's rare to experience someone who can shift and stretch people's deeply held paradigms in just a few moments, but Casey is one of those beings.

Content Notes: Settler colonization, Indigenous genocide, kidnapping of Indigenous youth

On the cultural and human genocide experienced by the Ponca Nation

I live in North Central Oklahoma, with my people, the Ponca. We are originally from Ni'bthaska, the area now called Nebraska. Our Trail of Tears happened in the late 1800s, when my grandpa was forced to walk nearly 680 miles with his parents and siblings and aunties and uncles. He was only six years old.

Over the course of history, we have entered five treaties with the United States as a sovereign people, as a nation of people with our own societal norms, religion, politics, and ways of being. In the course of those five treaties, we ceded over 2.5 million acres to the United States in order to retain a small area along what's now known as the Missouri River. During the treaty of 1868, they "accidentally" gave this territory to the Lakota to try to have access to the Black Hills and other railroad crossings. That's when my people were removed.

If we follow history, and it is a hidden history that is whispered about in our world and lives in our heart, it is one of dispossession and genocide that has never been fully recognized. In the late 1700s, in an act that I would call the first biological warfare tactic, smallpox blankets were introduced to us, and our tribe was reduced to around 200 people. Of the thirty-nine federally recognized tribes in Oklahoma, only three are Indigenous to this land; everyone else is here from the Trail of Tears. Thousands, millions of us died.

After genocide and forced removal, my mother, who was born in 1914, was among the first generation to be born in captivity on a reservation that was really nothing more than a prisoner-of-war camp. We were killed if we left the borders of the reservation. My mother was literally kidnapped into boarding school, where she and the rest of her generation were taken nine months out of the year from ages six through sixteen. Their style of clothing was replaced, their hair was cut, kerosene was poured on their skin to so-called rid them of whatever they thought was on them. Our language was stripped away. It was illegal to practice our religion. And so we had a generation that didn't fit in well at home, with the old ways that were still in their heart and genetic memories.

Around the same time, the Dawes Act was created to divide reservations into so-called allotments that were given to the individual.[1] Prior to that, we had always held land communally. Everyone took care of everyone. We had no jails. We had no old folks home. We never even had a word for homelessness. We did not use pieces of paper with writings on them to dispossess others. We didn't deal with ageism or sexism. Those were not in our societal norms. And then square buildings were erected to replace our circle way of life, where all was reciprocal.

The Bureau of Indian Affairs (BIA) was also created under the Department of War. It still exists, though is now under the Department of the Interior. The BIA forced us to certify the degree of Indian blood we had in order to be enrolled in our tribes, as if one can count how many cells are Ponca, or whatever nation, and how many cells are not.[2] And even if that concept is valid, what does it have to do with our spirit and heart and connection to our people? The BIA is akin to having a power of attorney over you from the time you were born. I was watching *Good Morning America* this morning and they were talking about Britney Spears and her father. I was thinking, *Okay, I get that*, because that's where we still are. It is an odd concept that has not only allowed but actually encouraged fossil fuel companies to come into our territory in the name of economic development.

I provide this context because it's important to understand part of what got us here.

On the environmental genocide experienced by the Ponca Nation

At the time that our people were forced into this cubicle reservation in the late 1800s, we had to leave behind our seeds, we had to leave behind our hunting and gardening equipment, everything. At the same time, when we first arrived here, we could still farm a bit. We could still hunt for the deer and fish. The water was still clear.

Since that period of time, Oklahoma has designated itself as an oil and gas state. They look for what they call resources, which we call sources of life. Polluters like ConocoPhillips, now known as Phillips 66, built a giant refinery and underground storage tanks, pipelines, injection wells, fracking sites, and one of the largest oil tank batteries in the country. They use all the water and poison the freshwater. All our water wells are shut down. There has been so much fracking activity that we have had 10,000 earthquakes here within a seven-year period.[3] Manmade earthquakes that literally created new faults! And the earthquakes are very shallow; while a typical earthquake in California might be 20 or 25 miles under the earth, here they are between 0.2 miles under the earth to maybe a mile. And we live on a spiderweb of pipelines that were not built for earthquakes, that are being destroyed with each tremor.

The city dump was placed across from our historic cemetery. A leachate pond sits across the street on the banks of the Salt Fork Arkansas River that runs through our community. A drop of that water will kill a human being. There is a coal-fired power plant about five miles from my home that still blows coal dust over the Otoe-Missouria Tribe's headquarters and housing. There is a facility leaking mercury and cadmium on the banks of the Arkansas River that runs by our home as well. Farmers and ranchers all around us are putting Roundup and various other kinds of poisons into the earth here. We can't grow organic food within about a twelve-mile radius of our tribal headquarters because everything is so polluted.

There is not one family among us that has not been multiply impacted by cancer. Because of the level of toxic pollution that we are all dealing with, when people go to the clinic here, which is substandard healthcare anyways, they find that they rarely have just one source of cancer but instead have multiple sources of cancer throughout their body when it is finally discovered. Babies are born with cancers here.

The pollution also causes autoimmune diseases. We have breathing problems that manifest in our babies as young as months old. All my grandchildren have had asthma. One of them was in the hospital every year up until he was fourteen, and his asthma still rages out of control. That is the norm here.

Our people are impacted by diabetes as a result of forced removal and the forced feeding of commodities that contain flour, sugar, white rice, grease, and dairy products, none of which were ever part of our diet.

Because we are poor—we probably have 65 percent unemployment here—we also live in multigenerational families. So the Covid-19 pandemic has wreaked havoc on our community and neighboring communities. One family had twenty-two family members fall sick with the virus. Just in our small tribe, we've lost a dozen or more wisdom keepers.

As for climate change, we're living in a moment of climate change as we speak. We are currently experiencing the polar vortex that has reached all the way down to Houston.[4] We have had subzero temperatures for days when our average this time of year is in the low fifties, and we're expected to have twenty inches of snow here very soon. This is the jet stream wildly careening as it never has before.

We were told by our medicine people to expect this, that Mother Earth had to purify herself because of what humans have done, and that it would come in this particular form. So when hurricanes and tornadoes started getting bigger and more frequent, when there was ice instead of snow, when we had more floods and worse droughts, we knew that it was coming and what it was about. We started to try and get people's attention back in the 1990s, when my friend Faith Gemmill began talking about permafrost melting in the Arctic Circle and how her people were going to have to learn how to garden for the first time in history. We still have not been able to get the media and others to truly pay attention.

There is no predicting what our Mother Earth must do in order to purify. Humans may get past this point, but they better listen now, and they better follow the Indigenous teachings to understand how we can interpret these patterns and preempt them.

On why Casey's an activist

Why do I breathe? Why do I eat? Why do I bear children? Why do I laugh? Why do I cry? Because that's a portion of being a human. That is the way that humans are supposed to be. It doesn't seem as if there's a why. You do it because there are generations to come that deserve the same exact thing. They deserve sacred water that will nourish their spirit as well. They deserve a clean breath of air.

When our children were born, I remember that first breath. It was like a gift from the winds. Where does that come from? Science will tell you it comes from the green things, those with roots. They breathe a certain way, and we breathe back. There's a reciprocal understanding of all that is. So I'm simply being reciprocal.

My family has always been what's now termed "activist." My brother, Carter Camp, was the leader of the American Indian Movement and later the Trail of Broken Treaties.[5] He partnered with Chicanos who were discovering their connections with their true ways and with the Black Panthers who were doing the same. We lived in a farm labor camp out in the San Joaquin Valley when I was around eleven or twelve, so we got to know the people we ended up marching with later, like Cesar Chavez. As we moved through years of civil unrest in the 1960s and 1970s, our family focused on reviving the cultural awareness and pride that would steer us back from Western society value systems and back into our value systems. In the 1980s we were reclaiming our religion and sovereignty and bringing awareness into other people's lives. And then there was a push by the Department of Energy to use our territories to dump hazardous waste, nuclear waste, and trash—all the things that nobody wanted in their backyard.[6] And so that brought another focus of activism for us, which is today called environmentalism. Really it is just a survival tactic for the human race.

On the power of Rights of Nature

Human beings have forgotten that without the plant life, without the water, without the air, without the fire, without the four-legs, without the fins, without the wings, we do not exist. Rights of Nature laws are another tool that we can use to stop the genocide of not only ourselves but all our relations.

When we talk about the Rights of Nature movement, it's important to recognize that we're not these big protectors of nature; we are nature protecting itself. And we are not giving nature rights. We are recognizing that those are the rights that are inherent, that the Great Mystery put into place. If you really internalize that understanding, then Rights of Nature is a natural way to hold those who are harming all of us, nature, accountable. It means that there is some way to hold the guilty ones culpable and to stop them. The purpose of Rights of Nature is to reestablish and realign natural law with human law. It's that simple.

RESILIENCE TOOL SPOTLIGHT

Rights of Nature

Rights of Nature, also called Rights of Mother Earth or Pachamama, is a budding legal movement, doctrine, and paradigm shift predicated upon an Indigenous worldview that humans live within, not above, the natural world. According to the Rights of Nature doctrine, ecosystems and the life forms supported within them have an inherent right to exist, flourish, and generate their natural capacities. If these rights are threatened or violated, then the ecosystem has the right to defend itself in a court of law from the source of harm, such as an extractive development project or even climate change. While some people are taken aback by the idea of granting legal personhood status to an ecosystem, it's important to remember that the US Supreme Court extended legal personhood status to private corporations in 1886. And as Casey reminds us, the fundamental rights of the **more-than-human** world are intrinsic whether or not they are recognized in courts of law.

It's long been recognized that all human rights, including the physical, mental, emotional, and spiritual well-being of human communities, are derived from the rights of nature, as nature is the foundation of life itself. In fact, it is the concept of nature as separate from humans, and as a thing that can be commodified, controlled, and dominated, that is relatively new. This worldview emerged as a central tenet of capitalism and colonization in the fifteenth century; early capitalists knew that people would not feel comfortable with the plunder of "a world where everything was alive and pulsing with spirit and agency, where all beings were regarded as subjects in their own right,"[7] and so they strategized with philosophers, the Church, and scientists to shift how people related to the world around them. As activist and educator Gopal Dayaneni explained in a workshop about rights-based organizing, "Rights are a new idea; we didn't *need* rights until we had entire economic systems predicated upon violence and the infringement of rights."[8]

The Rights of Nature legal movement is a promising alternative to the regulation-based framework that has undergirded agencies like the US Environmental Protection Agency, which has overseen a rapid decline in the country's ecological health since it was formed in 1970. As Shannon Biggs, Mari Margil, and other Rights of Nature activists have illuminated, environmental regulation has not only been futile in preventing or ameliorating ecological catastrophe, it has actually codified and enforced the continuation of pollution and destruction. Rights-based frameworks, on the other hand, are proactive. Rather than test the bounds of harm that ecosystems and their inhabitants can endure, Rights of Nature laws seek to repair and avoid harm altogether.

Rights of Nature laws are therefore a potentially powerful tool to oppose fossil fuel development projects that both destroy surrounding ecosystems and exacerbate the climate crisis. In a system where the elevation of corporate rights over communities and natural systems is constitutionally embedded, Rights of Nature laws may be key in helping communities resist projects that threaten their forests, rivers, mountains, deserts, lakes, and their own well-being. It may be a particularly potent tool for tribal nations across Turtle Island to safeguard their subsistence practices, sacred sites, and sovereignty on their traditional lands. The Rights of Nature framework may also be instrumental in instigating a larger economic, legal, and political paradigm shift predicated upon a recognition of our interdependence with the more-than-human world, or as Casey would say, "the oneness of all and the allness of one."

Thrillingly, the Rights of Nature movement has been gaining traction around the world for the past couple of decades. In Bolivia, Ecuador, and Panama, the rights of Mother Earth have officially been recognized. The Amazon River in Brazil, Whanganui River in New Zealand, Ganges and Yamuna Rivers in India, and all rivers in Bangladesh have been granted legal standing. Several tribes, like the Ponca, Ho-Chunk, and White Earth Nations, have passed laws enshrining the rights of nature, in part to resist fossil fuel development and other harmful projects, to mixed results. Several municipalities across the so-called United States have taken a similar tact, though most have been met with resistance at the state and federal levels. In Orange County, Florida, for instance, 89 percent of constituents in the 1.4-million-person county voted in favor of a Rights of Nature amendment, only to find out that a twenty-nine-word clause was snuck into a 111-page state bill weeks prior that would prohibit local governments from recognizing the legal rights of nature. There are glimmers of hope, though. After the small township of Grant, Pennsylvania, was sued by the Pennsylvania Department of Environmental Protection (DEP) and Pennsylvania General Energy Company for resisting an injection well via a Rights of Nature amendment in their town charter, Pennsylvania DEP revoked the permit for the injection well. These moments of success set a profound precedent for what's possible.

To get involved with the Rights of Nature movement in your community:

- **Research prompts:** Are there any Rights of Nature efforts emerging in your region? Are any Rights of Nature cases around the world at a critical juncture and calling for resources or other forms of solidarity? Pick a Rights of Nature case to explore in more depth. In what ways has the effort been met with resistance? In what ways has the strategy been successful or could it pave the way for success?
- **Journal prompts:** How is your well-being and the well-being of your community bound up in the ability of local, regional, and/or global ecosystems to exist, flourish, and regenerate? What does interdependence with the more-than-human world look and feel like to you? In what ways could you honor this interdependence in your daily life?
- **If you're already engaged with your community or a local movement group:** Start a discussion about the Rights of Nature movement.

Host a screening of the documentary *Invisible Hand*. Gather friends to join a virtual Democracy School session through the Community Environmental Legal Defense Fund (CELDF). Organizations like Movement Rights will partner with your group to host an interactive, rights-based organizing workshop or all-day training. If you're a tribal member, reach out to Movement Rights to learn more about their Intertribal Rights of Nature Forum offering.

- **If you have a background in law:** Get trained so that you can offer pro bono assistance to the Rights of Nature movement. Look to organizations like the Earth Law Center and CELDF for trainings, learning resources, and networks to plug into.
- **If the ecological well-being of your community has been impacted or is being threatened:** Consider organizing with other community members to get a Rights of Nature amendment on the ballot in your municipality. Several organizations have programs in place specifically to assist communities in Rights of Nature efforts, so lean on their support and resources to guide you through the process. CELDF, for instance, offers pro bono and low-cost legal assistance and organizing assistance to community groups interested in advancing rights-based legislation.
- **If you have access to financial resources:** Lend support to Rights of Nature efforts around the world. Move resources directly to the communities who are leading the way, or set up recurring donations to organizations like CELDF, Movement Rights, WECAN, the Bioneers Indigeneity Program, or Global Alliance for the Rights of Nature.
- **If you have a passion for the written word or content creation:** Raise awareness about the Rights of Nature movement or a particular case. Write an opinion editorial for your local newspaper, or share information with your social media network.
- **General best practices:** Rights of Nature advocate Mari Margil says, "Protections are best shaped by the communities that wish to establish

them, whether they're concerned about drinking water, oil spills, or the survival of a certain plant or animal." No degree or professional experience is necessary to get involved in the Rights of Nature movement; your lived expertise is more than enough.

VICTORIA MONTAÑO

THEY/SHE/HE

"Land Back is possible, even in an urban setting"

Victoria Montaño, also called Victor, Vick, or V, is a first-generation, two-spirit, Yo'eme and Mexikah person born and raised in the Lisjan Ohlone village of Huichun, presently known as East Oakland, California. For the last several years, Vick has been a dedicated member of the Sogorea Te' Land Trust, an urban, Indigenous women–led land trust founded in 2015 by Corrina Gould and Johnella Rose after they had been organizing against the desecration of sacred Lisjan Ohlone sites for decades. Today, Sogorea Te' Land Trust is working to establish a land base for Lisjan Ohlone people, who were dispossessed from their ancestral homelands by colonizers hundreds of years ago and today are disproportionately impacted by housing, food, and income insecurity along with other forms of political, economic, and cultural marginalization.[1] The Sogorea Te' community is currently lovingly restoring a small but growing constellation of returned land throughout the so-called East Bay. They are also revitalizing culture through song and ceremony, language classes, community events, and more. They have been instrumental in facilitating Indigenous land return to Indigenous hands far beyond the so-called Bay Area; in the last few years, their work has sparked conversations and action in many of the environmental, justice, and philanthropy spaces that I'm a part of.

As a member of Sogorea Te's land team, Vick stewards the land and waterways and grows medicinal, edible, and sacred plants. They provide spiritual, physical, and emotional care for their community by cultivating and distributing traditional medicine. They also use their gifts and passion as a digital and visual artist to powerfully communicate timely, resonant messages and to act in solidarity with other Indigenous movements. In fact, when Niria Alicia introduced me to Vick, I soon realized that I had already experienced many of their pieces via Instagram and posters and murals I'd seen throughout the Bay Area. Vick's work has reached far and wide! It is inspired by and in honor of their ancestors, Indigenous women and leaders, first-generation, Spanish-speaking, and queer folks, the elements, the plants, and the next generation. In all that Vick does, they move through the world with tremendous thoughtfulness and intention.

Content Notes: Sexual violence, structural and interpersonal racism, anti-queer discrimination, settler colonialism

On their journey into the Land Back movement

I always struggled with my identity growing up. When I went back to Mexico for the first time, visiting the place my parents immigrated from in the late 1980s, I felt like everyone looked at me like I was an American. But in so-called East Oakland, I'm just a Brown person. And in the Indigenous community, I struggled with a lot of close friends questioning my Indigeneity. In high school I was asked about my blood quantum and my tribal identification card. I didn't know or have any of that.

I'm still trying to learn my traditions. Because I'm two-spirit and I'm queer, it's particularly difficult. Not knowing how people are going to respond to me, it makes me nervous to reintroduce myself to a community that I didn't grow up around. As I've gotten older, I've separated myself from some of the things that were said to me, and I now recognize that it is an Indigenous experience to have these identity issues and to have other Indigenous people question your Indigeneity. It's all part of the Indigenous struggle. At the end of the day, these are false, manmade borders and the focus should instead be on collectively trying to help Mother Earth heal. The work I do today reflects that.

A few years ago, Corrina Gould, the codirector of the Sogorea Te' Land Trust, gave me an opportunity to be a part of this dream. Sogorea Te' Land Trust emerged from the work that Corrina Gould and Johnella Rose have been doing for more than twenty years, leading shellmound walks, trying to gain federal recognition for the Lisjan Ohlone, and occupying a sacred site in Vallejo.[2] If it weren't for the work that they've done, most people wouldn't know that, in this urban setting, we're still on Indigenous land. Years ago, Corrina was invited to a land trust conference and noticed that it was a boys' club. After reflecting on it and discussing with other community members what healing could look like for the Lisjan Ohlone, the idea of the land trust was born. The whole goal of Sogorea Te' is to provide space for our community to heal, to bring back language, ceremonies, and songs, and to have a place for Ohlone babies to access these traditions.

We're in the midst of a huge movement, and a lot of people are looking at Sogorea Te', an urban, Indigenous women–led land trust, for guidance. So many people still think about Land Back as a hashtag, but more and more people are seeing that this work is possible, even in an urban setting. We're seeing land being returned to our relatives. Sometimes it's not the best land. Some of the pieces of land that have been returned to us are trashed. We've had to really tend to it. But we've still been able to create paradise in East Oakland. I want that to give people hope.

We're able to do this work and create these spaces thanks to people who have had a mindset shift and want to pay the Shuumi (which means "gift" in Chochenyo) land tax. It allows Sogorea Te' to pay thirteen team members. Our ancestors have been doing this free labor for way too long. This work can be hard on our mental health and physical bodies, and we deserve to be paid for it. It's still kind of a shock that people pay this tax, but I hope people realize that there are other ways to practice rematriation, too, whether that's returning land or handing over housing or contributing skills to the movement.

On the importance of traditional medicine restoration, cultivation, and respect

As a member of the land crew with the Lisjan Ohlone, I've been reintroduced to the land and soil. The land has transformed me. It has allowed me to reawaken and to reconnect to the traditional, ceremonial medicines, not just of native California, but also of my own ancestral lands.

I recently had an elder tell me that all the things that I've experienced in life have prepared me to take on this work. I always like to say that the ancestors selected me for this. Losing a sister to suicide and having Indigenous friends murdered, I do this work in a way that's deeply connected to spirituality and the spirit world. The soil, land, and plants have spirits. Knowing that they're interconnected with the spirit world, it's helped me accept that death isn't the end. I honor the ancestors by doing this work. I use my experiences to feel more grounded in the land.

For me, success looks like having land to grow traditional plants, care for the soil, watch plants grow, harvest them, collect their seeds, and hand seeds over to another Indigenous person, so that they can plant those seeds in their own garden and continue to grow that medicine. It looks like passing down knowledge and making traditional medicine for our community, to allow us to step away from pharmaceuticals and their side effects, and to provide options beyond a healthcare system that many of our community members don't have access to. During fire season, when smoke lingers in the Bay Area, and many of our community members are deeply affected, it just reiterates how important this work is. We're successful when we can provide healing salves to our elders, who oftentimes had jobs that did permanent damage to their bodies. Or when we're able to make tinctures from California poppies and chamomile and mugwort, which calm the nervous system down and are good for anxiety, because we want to care for our community members' emotional health as well as their physical health. When we're able to grow white sage and tobacco and harvest and share those seeds, we're able to provide spiritual medicine to our community, too.[3]

On the power of art to call out, call in, and invite folks into the movement

I'm a multimedia artist and I like to get my messages out through my art. I recently designed a graphic that says *Land acknowledgments are not reparations*, because it's not a trend, it's an action. Another graphic that I'm really proud of says *Speak about the Ohlone people in the present tense*, which I'm heavily committed to doing as long as I'm alive. I've been very blessed to be a part of a lot of art builds with movements like Mauna Kea, Run4Salmon, and, most recently, a design to bring awareness to the West Berkeley Shellmound.

There have been times when I've been sitting at tables with very intellectual people, like professors and scientists, and I've felt really insecure. Then I realized that my voice is important, and I use my voice through the artwork that I put out. I also realized that this is how I'm supposed to reach the people that I reach. If I tried hard to speak this language, and use these ten-letter words, I wouldn't be able to reach the people that I grew up with in the neighborhoods that I grew up in. So fortunately I've separated myself from that insecurity.

I still deserve a seat at the table where these conversations are happening. I remind myself that Corrina and Johnella trust me and that there's a reason why I'm here, involved in this work. It goes back to that feeling; I was selected by the ancestors..

RESILIENCE TOOL SPOTLIGHT

Land Back

While the concept of Land Back has existed since white settlers first violently stole Indigenous lands, the contemporary Land Back movement began over a decade ago as culture bearers and artisans disseminated the idea through art. The movement has recently gained visibility and traction thanks to land and water defenders, digital activists, and organizations like NDN Collective, who launched the LANDBACK campaign on Indigenous Peoples' Day 2020. Today's movement builds upon many generations of effort across Turtle Island and beyond to address the roots of colonization and reclaim stolen Indigenous lands.

The Land Back movement is not a monolith. For some, Land Back is literal; as Lakota matriarch Madonna Thunder Hawk says, "The only reparation for land is land." For others, Land Back is primarily about sovereignty and cultural reclamation. Krystal Two Bulls, the director of NDN Collective's LANDBACK campaign, offers specific demands that encompass both: (1) to return all public lands back to Indigenous hands, (2) to dismantle the white-supremacist structures that keep Indigenous peoples oppressed, (3) to defund and dismantle the systems that disconnect Indigenous folks from the stewardship of land, including the police, military, and border patrol, and (4) to shift policy from consultation to free, prior, and informed consent.

While there's still a long way to go until the demands outlined by Two Bulls are met, there are countless examples of land and water return unfolding around the country, from a farmer in so-called Nebraska signing a deed returning ancestral tribal land back to the Ponca Tribe to the Eureka City Council voting unanimously to transfer ownership of the sacred Duluwat Island to the Wiyot Tribe. There is no singular, straightforward land return process, but more and more replicable Land Back models are emerging, including community land trusts, deed transfers, dam removal, co-management, and voluntary land taxes, among others.

The Land Back movement is increasingly being recognized as a cornerstone of environmental and climate justice, particularly since Indigenous peoples are proven stewards of lands, waters, and skies; Indigenous peoples comprise just 6 percent of the world's population, but Indigenous-managed lands protect approximately 80 percent of the world's biodiversity.[4] To prevent future ecological harm and repair the harm that has been done, it's imperative that relationships between Indigenous peoples and their ancestral homelands are restored.

To support and accelerate the Land Back movement as a settler on stolen lands:

- **Research prompts:** If you're not currently residing upon the traditional lands of your ancestors, whose land are you on? What is the history of the land? Which traditional languages were/are spoken? What treaties were established by Indigenous tribes and settler governments? What was originally specified in the treaties, and have they since been honored by the settler state? How are Indigenous peoples in your region asking you to engage in the healing and repair of broken responsibilities with the land and their people? The Native Land app and website (www.native-land.ca) are great resources to begin your exploration, but keep in mind that information is crowdsourced and should serve as a starting place.
- **If you have access to financial resources:** If you're currently residing someplace other than your ancestral lands, make a monthly or annual contribution to the Indigenous communities that originally stewarded

the land (if they have consented to receiving monetary donations). If you're a non-Indigenous business owner or leader of a school, group, or organization, an institutional contribution should be added to your annual expenses. This payment is sometimes referred to as a land tax, real rent, honor tax, or Shuumi. Sogorea Te' Land Trust's land tax calculator is a great tool to utilize if you're unsure how much to give. This payment is a small way to acknowledge the territory's history, repair relationships, and support the return and restoration of Indigenous land to Indigenous people. Make sure to ask your local government to pay their land tax, too.[5]

- **If you have access to land:** Invite local tribes to utilize the land in whatever ways make sense for them, such as holding ceremony, growing traditional foods and medicines, gathering occasionally, and so on. If you own stolen lands, consider returning the land to its original stewards. Since the concept of individuals legally transferring land to Indigenous peoples is relatively new, there isn't a singular, straightforward process. Instead there are myriad potential pathways and land ownership structures to explore. Begin the process by building relationships with Indigenous peoples in your region. Go slow, be patient, and honor complexity as it arises. Explore possibilities together.
- **If you have access to time:** Contribute your time, talents, and solidarity to Indigenous-led efforts to protect and reclaim their ancestral lands and waters. This may look like contacting your local and state representatives about projects that violate treaties and threaten Indigenous lifeways, like river dams and fossil fuel infrastructure, reaching out to federal representatives to let them know that you support the return of national public lands to Indigenous communities, or calling for your local land trust to return its land to its original stewards. It also might look like making art for protests, sending supplies to the front lines, gathering friends to show up for sit-ins and marches, offering pro bono legal services to help advance legal action, and on and on. There are infinite ways to get involved.

- **If you have access to a robust community network:** The previous action items are best done together. Gather friends, neighbors, and family members for an evening of good food, discussion, paying land taxes, contacting representatives, and/or organizing to support Indigenous-led efforts.
- **General best practices:** Follow the lead of Indigenous peoples in your area. Focus on building relationships. Show up with humility, patience, and consistency.

Ecological Restoration

Ecological repair is a powerful tool in bringing weather patterns back toward long-standing norms and enabling ecosystems to more effectively buffer climate impacts, sequester carbon, keep our human and more-than-human kin safe and nourished, and so much more. These essays are from ecologists, planners, earth workers, and educators who are illuminating the harm caused by attempts to manipulate and dominate the natural world and who are lovingly and diligently working to restore waters and lands, as well as the ecological stewardship practices that were disrupted by settler colonization.

HEATHER ROSENBERG

SHE/HER/HERS

"So-called natural disasters and social injustices are not two different sets of issues"

I was eager to include Heather Rosenberg's perspective because she has been one of the foremost leaders in ensuring that the built environment of tomorrow will keep communities safe in an age of more frequent, heightened disasters. Heather has been shaping the field of sustainable and resilient infrastructure for the last twenty years, and today she serves as the leader of the Americas Resilience Discipline at Arup, a global, employee-owned engineering and planning consulting firm. An ecologist by training, Heather brings a keen, **systems thinking** perspective to the role, where she convenes myriad technical experts, from structural engineers and urban planners to risk assessors and community organizers, in order to mitigate potential risk in the most holistic manner possible. While the scope of Heather's role is global, the lifelong Angeleno spends most of her time developing projects local to Los Angeles, capturing lessons learned, helping determine which pieces are scalable and replicable, and communicating that information to communities elsewhere.

Additionally, Heather was appointed by the Los Angeles mayor to serve on the city's Innovation and Performance Commission and the Resilience Steering Committee. She worked with the Global Real Estate Sustainability Benchmark (GRESB) to develop a Resilience Module for real estate investment companies and funds (a framework that I myself utilized while working as a sustainability director for a real estate firm; GRESB has become the industry gold standard for environmental, social, and governance efforts). She was awarded a prestigious US Green Building Council Ginsberg Fellowship, founded the Building

Resilience-LA program to expand physical resilience efforts across the city, and authored several accessible and instrumental primers on resilience and social equity in the built environment. She also founded the Building Resilience Network, a strategic planning and capacity-building firm. Heather deftly dances between public, private, nonprofit, and organizing spaces to construct a more just and sustainable society for all people.

Content Notes: Slavery, settler colonization

On the necessity of solving multiple problems at once

This country was built upon an ethos of dominating nature. I think that the Los Angeles River is a great example of this. The LA River project of the 1930s was designed to solve a very real, very significant problem: there was massive, devastating flooding along the river that went through the city and killed people on a regular basis. So the US Army Corps of Engineers solved that problem really well by deciding that the river shouldn't be allowed to move and should instead be channeled through a very hard, concrete system.[1] While this approach effectively addressed the flooding, it didn't take into account all the other challenges of the community or any of the other functions of what a river should be. Today the LA River provides no open space, recreation, connection with nature, habitat, filtration to capture and restore local water sources,[2] or mental or public health benefits—nothing. It is a nasty, concrete gash in the landscape.

And now there's an opportunity to rethink it. I actually live near the LA River and see some of that work happening; people are cutting out the concrete, building parks, and pulling in natural solutions that are much more compatible with the larger ecosystem. There's a concerted effort to consider all the functions of a river and the surrounding areas. And I think we need a similar approach to all the problems that we're dealing with. We cannot afford to solve one problem at a time. Our problems are too big. We need to solve them all at the same time.

Los Angeles is, in a sense, a microcosm of everything that can go wrong. We experience pretty much every type of disaster that FEMA recognizes. We have a very diverse and complicated risk profile, with everything from tsunamis and earthquakes to fires, floods, zombies, and sharknadoes; there's a reason why so many disaster movies are set in Los Angeles! And at the same time, Los Angeles has an enormous population of unhoused folks.[3] There is huge racial and income disparity.[4] And all these things have to be thought of together. Even with all the aforementioned disasters, heat continues to be the biggest killer of any hazard here, and that's largely because so many folks are refused safe working conditions or do not have access to adequate cooling and housing.[5] So-called natural disasters and social injustices are not two different sets of issues. It's one set of issues.

So as we consider possible solutions, we need to be very clear about what our goals are. What is it that we're trying to achieve with the systems that we're designing? Our priorities are manifested in the built environment, and right now our priority should be to serve the entirety of our communities under a vast range of conditions, knowing that disruption is coming.

On designing just and sustainable societies

I got into the sustainability world right as the U.S. Green Building Council's LEED (Leadership in Energy and Environmental Design) rating system was being launched. I was able to see firsthand how market-based tools can have a real impact, and I was very inspired by the practical side of that. However, as I got deeper into it and worked on green building for many years, I got really frustrated. I'm an ecologist by training. I look at the world in terms of systems. But when you look at the field of green building, and even sustainability as a discipline, it can be very mechanistic; it's generally about optimizing for the best possible situation for building performance without considering people. How green is your building, though, if it's displacing a community? How green is your building if it's exploiting its workers? When I started to ask myself where the leverage points are to get green building stakeholders to care about these questions, I didn't find very many. In the context of market-based solutions, it's really hard to talk seriously about social justice. Or it's easy to talk about, and nobody will listen.

The reality, though, is that you can't talk about green building or community resilience in the United States without first acknowledging that we are a country built upon slave labor on stolen land. We must understand deeply that we're working with systems that are fundamentally unjust. We need to recognize that social disparities are institutionalized, not accidental, and figure out where the forces are that pushed us in those directions, so that we can work to repair and heal those roots.

Right now, I get to work on a project with a group called Designing Justice + Designing Spaces where we're looking at possible community-based approaches to alternatives to incarceration in Los Angeles.[6] Drawing from the Los Angeles County Alternatives to Incarceration Working Group, we're thinking about what communities of care could look like and how they might be designed. Projects like these are so rich and exciting because they get to the heart of what it means to create a just and sustainable society. If we were building a just and sustainable society from scratch, we wouldn't start with the militarization of police or an industrialized prison system. So where could we allocate that funding instead? What would we want to build and create?

On reframing resilience

First of all, it's important to recognize that people are already very resilient. The issue is generally not a lack of resilience, but rather that people are already using all of their adaptive capacity. It's important that we're not asking people working three jobs, trying to get their children through school, fighting to keep the lights on, to be *more* resilient. Instead of asking folks to cope with even more, the real goal needs to be alleviating underlying stressors so that they have more capacity to draw from.

If resilience is like a rubber band, and it is already stretched as far as it can go, it's crucial that we're not trying to get more out of our rubber bands but instead finding ways to demand less from them. Whether we're talking about poverty, racism, crumbling infrastructure, under-supported schools, under-resourced streets, or over-policing, these things are stressful and draining in a whole variety of ways. The more that we can address those root issues, the greater capacity folks will have to effectively deal with disruptions when they occur.

Therefore, resilient communities are equitable, where everybody has their basic needs met. They are connected, where people have a strong sense of relationship to each other, to the place they live, and to purpose. And they are empowered, where community members have access to power and can participate in the decisions that will impact their lives. Equity, connection, and empowerment form the bedrock of community resilience, and if you have those things, then you can get into the more granular aspects of physical resilience to disasters and focus on lifting up the solutions that communities already possess.

RESILIENCE TOOL SPOTLIGHT

River Restoration

River restoration is an umbrella term for efforts that aim to improve the ecological processes and integrity of degraded waterways and reinstate more-than-human members of the watershed who have been harmed, displaced, or eradicated. River restoration practices vary widely depending on the state of the watershed, goals for restoration, and resources available. Many river restoration projects include dam removal, which can restore natural fish migration patterns and allow populations of fish, particularly those that migrate up rivers from the sea to spawn, to bounce back. Dam removal projects also help improve water quality, reduce flood risk, restore natural river flows, and recharge aquifers. When rivers can carry sediment all the way to the ocean instead of remaining trapped behind a dam, the deposited sediment is able to reinforce riverbanks and bolster river deltas, therefore helping mitigate coastal and river flooding. Reintroducing native plants and riparian forests along riverbanks can keep river temperatures cool, soak up rains, mitigate floods, prevent erosion, and bring back and support members of the river ecosystem. As the climate crisis intensifies periods of drought, heavy precipitation, and extreme heat, it is crucial that river ecosystems can effectively absorb some of those shocks.

The original white settlers of the so-called United States failed to understand how their own well-being was intertwined with the well-being of the rivers, and they had little regard for the countless ecological, cultural, and spiritual roles that rivers and their inhabitants possess. Fur trappers devastated beaver populations for their pelts, which were fashionable in Europe, and "the disappearance of beavers dried up wetlands and meadows, hastened erosion, altered the course of countless streams, and imperiled water-loving fish, fowl, and amphibians—an aquatic Dust Bowl."[7] Rivers were utilized as industrial sewers and shipping lanes and manipulated to power machinery and supply water to faraway communities. They were dammed, diverted, buried, polluted, privatized, and commodified.

Today, with more than two million dams, culverts, and other barriers across the so-called United States,[8] less than 2 percent of the three million miles of rivers and streams across the country are free-flowing and unobstructed.[9] However, the tide is finally turning for many watersheds. Most dams across the country have now surpassed their intended fifty-year lifespans, making them a danger to their surrounding communities (particularly with increasingly erratic rainfall and floods). Largely because it can be less expensive to remove old dams than bring them up to current standards, over 700 dams have been taken down in the past decade.[10] The effects have been overwhelmingly positive.

Additionally, there have been several recent victories in Indigenous-led fights to remove major dams and bring rivers back into balance. The Elwha Dam removal, the largest dam removal project in history, was completed in 2014 after over a century of protest from the Lower Elwha Klallam people, whose treaty rights were violated by the blocked passage of fish. Further south, the Yurok, Karuk, and Klamath tribes, who have relied on salmon for food, ceremony, and lifeways for thousands of years, led the charge to remove several dams in the Klamath River. Four dams along the river are set to be removed by 2023, restoring salmon access to more than 400 miles of habitat. On the opposite coast of Turtle Island, the Penobscot Nation formed a trust with other nonprofits to buy up hydroelectric dams blocking the passage of anadromous fish. They found a way to maintain production of hydroelectric power across the river while removing two of the dams and building a fish passage around a third.

Large-scale river restoration and dam removal projects are not straightforward nor easy, though. There are often numerous communities weighing in: engineers, scientists, conservationists, farmers, recreationists, energy companies, local, state, and federal governments, and Indigenous nations. It's incredibly difficult to find alignment across all stakeholders. Particularly in densely populated and water-scarce regions, where water rights tend to be over-allocated, these processes can be decades long, expensive, and highly contentious. River restoration efforts can also be logistically challenging. For example, many communities have been built in areas where rivers used to flood or even flow, so restoring the integrity of a river could put homes and lives at risk. Additionally, bringing a river back into balance is a delicate dance and there are many unknowns; even reinstating native plants can throw off an ecosystem that has adjusted to their absence.

And yet, fighting for the health and sanctity of our rivers will always be worth it. As the Lakota protest anthem for Standing Rock declares, mní wičóni, *water is life*.

To get involved with the river restoration movement in your community:

- **Research prompts:** What are the names of the rivers and streams in your watershed? What can you learn about their histories? What can you learn about the current health and well-being of the nearby rivers? Are there any Indigenous-led river restoration efforts underway in your region? Which committees, councils, or task forces in your city and/or county address waterway issues, and do they have room for more members?
- **Journal prompts:** What does your relationship to nearby rivers and streams look, sound, smell, taste, feel like to you? What does *water is life* mean to you?
- **If you have the financial resources:** Support Indigenous-led efforts to protect and restore rivers and their more-than-human inhabitants, such as Run4Salmon's Salmon Run Fund, a Winnemem Wintu Tribe-led effort to return Chinook salmon to the McCloud River watershed.
- **If you have the passion and bandwidth:** Apply to join your city or county's committee, council, or task force that addresses waterway

issues, or attend the meetings where matters related to river infrastructure, river protection, or ecosystem restoration are on the agenda. There's always power in numbers, so organize friends, colleagues, and neighbors to show up and speak out together. Alternatively, if there are river restoration efforts already underway in your community, ask how you can be of support.

- **If you want another excuse to get outside:** Make a regular ritual out of cleaning up trash along the banks of a nearby stream or river. It's a meditative practice and a beautiful way to deepen your connection with your watershed. Bring gloves, trash bags, and a trash picker if you have one. If you're not already familiar with a waterway near you that could use some love, reach out to a local watershed association, environmental group, or stream watch group to see where they recommend going, or join a cleanup that they've organized.
- **General best practices:** As with all land and water restoration efforts on colonized lands, Indigenous folks should be front and center in efforts to restore their traditional waters. As Yurok fisherman and activist Samuel Gensaw III shared on the *For the Wild* podcast, "I believe if you are part of the conservation effort and that effort does not put Indigenous people at the forefront of the decision-making process, if that organization is not implementing Traditional Ecological Knowledge, and working with the Indigenous people of that area, you are a part of modern colonization."[11]

CATE MINGOYA-LAFORTUNE

SHE/HER/HERS

"This work is an opportunity to center the needs of the people who've had to live with the legacy of industrialization and an extractive economy"

Cate Mingoya-LaFortune is a biologist, city planner, student of history, organizer, and storyteller. When Cate and I spoke in the winter of 2021, she was working as the director of capacity building for Groundwork USA, a nonprofit working to transform structurally neglected urban spaces into hubs for resilience, connection, and well-being in a constellation of twenty-one cities across the country. In that role, Cate was tasked with strengthening the network's connective tissue, providing technical assistance to brownfield remediation projects, and leading their Climate Safe Neighborhood program, a partnership designed to help expose the persistent effects of housing discrimination on climate vulnerability while buttressing community resilience efforts. With each challenge, Cate worked hard to not only support impactful, community-driven solutions and cultivate people power, but also challenge the very power structures that created the problems in the first place. I was particularly excited to speak with Cate because her approach touched so many building blocks of resilience at once—bolstering community green space, creating areas for folks to gather and get to know one another, mitigating harmful contaminants that can be exacerbated by extreme weather, strengthening local organizing muscles, and so on. Today, Cate serves as Groundwork USA's national director of climate resilience and land use. In this capacity, she's focusing even more deeply on climate adaptation and working with residents to build their capacity to intervene in local systems for a more equitable distribution of climate adaptation resources.

Cate's work is informed by her own experience as a resident of historically redlined communities, growing up in Queens, New York, and raising a family in Somerville, Massachusetts. A former middle school science teacher in the Bronx (she still speaks with the ease, magnetism, and speed of a seasoned educator), Cate was inspired to pivot into environmental and housing justice spaces because she knows that youth deserve to grow and thrive in their own communities. She stays inspired by the desire for more green spaces for her daughter to play in and clean air for her to breathe.

Content Note: Structural racism and classism

On shifting the framing of environmental injustices

I grew up in the wonderful and diverse neighborhood of Bayside, Queens, New York, just south of Northern Boulevard. Its physical infrastructure historically supports a community of residents who are constantly in transition and are marginalized by larger systemic powers. In Bayside, we had one tree on the corner. In the summer, when our parents locked us out to play outside, we would beat the crap out of each other to sit under the shade of that one tree. We cared so much about that tree. It's still there and is still the only tree on that block.

Additionally, we were surrounded by contaminated sites and had no access to green spaces. When we wanted to play outside, we would crawl underneath a little space in the fence to get to a **brownfield site**, where we would then kick up and breathe in toxic contaminants. The combination of living near contaminated sites and across from a major thoroughfare meant that I, and many of my neighborhood friends, grew up with asthma.

I grew up being told, *It's your job to go get an education and leave this neighborhood*. There was an expectation that, to be successful and safe, and to have access to things that we didn't have in our neighborhood, we needed to move elsewhere. And that's what I did. I listened to all the well-meaning adults in my life, and I went to college and got a degree in biology, in part because it seemed like a viable path to getting out and staying out. After I graduated, I decided to go back to New

York and teach. It felt like the right next step; my mother had been an early childhood teacher, I had loved learning, and it also felt like I had gotten really lucky with the scholarships and educational opportunities that I had received.

When I was teaching in the South Bronx in East New York, in neighborhoods that looked a lot like the ones that I grew up in, I found myself saying the exact same thing that the adults in my life had said: *You need to get a good education, and then you need to go.* One day at dismissal, as the teachers headed to the train and the students peeled off to go to the housing projects, I really took in my surroundings. For a mile in each direction, all the eye could see were the really tall brick buildings with bars on the lower windows, doors that don't close, lots out front filled with burnt-out couches, playgrounds outside that aren't being utilized and lack green space entirely. At that moment, it struck me: we get to go home and leave this neighborhood, but our kids are stuck in a place that doesn't necessarily feel safe. And it's a place that has their family, their language, their culture, and their communities.

What I had been told, and what I was then mistakenly parroting, was that the only way to be safe and have access to clean air and economic opportunity was to leave. But that meant leaving behind language, family, and culture. It's a lot to look a person who's eleven or twelve or thirteen years old in the eyes and say, *Your community and family and relationships aren't good enough and you need to leave them behind. And when you leave, you can't really come back, because cities are constantly changing, and it may be too expensive to live here again.* It asks them to do something really painful. It also makes it seem like their communities are the source of the problem, when really it's a broader system that needs to be addressed. So instead of repeating that incredibly hurtful message, why aren't we focusing on building community power, transforming the built environment, and obtaining the resources that these communities have been intentionally denied for so long? Why aren't we shifting the system so that kids can stay with their families, language, and culture, explore and leave if they'd like, and be able to come back?

On the transformation of brownfield sites into green spaces

One of our main focuses at Groundwork USA is remediating brownfield sites and working with community to transform them into green spaces.[1] Brownfields are sites that are contaminated or expected to be contaminated. They are generally the legacy of industry. They didn't come out of nowhere; at one point in time, they were producing something that was useful and valuable, like asbestos for housing siding or lead paint. These industrial sites, oftentimes manufacturing factories, would be sited in places that have good access to transportation, like alongside rail yards that could be used to ship out their product, along with good access to sources of power, like river waterways.

This layout was really practical, but there were downsides for the people who worked at these factories and lived nearby, generally low-income folks and recent immigrants. For one, factories were oftentimes actively pumping contaminants both onto the land and into the adjoining water. Rail yards were also contaminants; the rail ties contaminated the land, while the train smoke was really damaging to long-term health. The rivers, which were great at producing power to run factory equipment, also tended to flood. Therefore, surrounding areas tended to be wet. When people think about flooding, they generally think about someone's car getting washed away dramatically down a street. However, there are many other insidious ways in which water leads to hurting folks, like damaging everything in the basement or sprouting mold in the walls. And if the water is contaminated, it contaminates the entire area when it floods.

Brownfield sites also add a certain level of blight to a community. Planners often talk about the **highest and best use** of a property, and that tends to be thought of exclusively in terms of economic value. Since brownfields are usually really expensive to clean up, they don't often reach the threshold of a developer getting a highest and best use for it. Therefore, a lot of neighborhoods with brownfields are dramatically underinvested in; they lack green space, new affordable housing and developments, opportunities for urban agriculture or recreation or communing with nature. The long-term effects of brownfield sites speak to a larger legacy of extraction; someone came, built, extracted resources from the community in the form of labor and clean air and water and soil, and then stripped the community of their ability to use the land easily and inexpensively for generations afterward.

My coworker Lesly Melendez, who is from the community and is now the Executive Director of Groundwork Lawrence, tells one of my favorite stories about the Spicket River in Lawrence, Massachusetts. There used to be a paper mill along the Spicket River that manufactured the glossy paper used in magazines. They used to dump the chemical residue from the manufacturing process into the rivers. If you talked to longtime residents, they would tell you that they could tell what color paper they were working on because the Spicket River would turn green, purple, red. The river was so toxic that the neighborhood's version of *Go f*ck yourself* was *Go fall in the Spicket.*

For so long, the Spicket River was really disgusting. It was filled with chemical residue and unwanted refrigerators and couches. Over a twenty-year period, Groundwork Lawrence was able to build the Spicket River Greenway, which is a 3.5-mile greenway programmed to meet people's psychological, physical, and environmental needs. You can physically engage with the space by walking, jogging, or biking along the greenway. There's an amphitheater used for recreational programming that doubles as a flood basin; when the river floods, the water flows into the amphitheater and stays away from the commercial and residential areas on the other side of the Greenway. A bit further inland, there are community gardens where folks connect with one another and can grow culturally complementary foods—the herbs and spices that connect them to home and allow them to continue cooking their family recipes. Over the marsh, there's a bird-watching platform where folks can watch birds or participate in an evening of tango dancing. Throughout the Greenway, the tree canopy has been densified to provide relief on hot, humid days.

The Spicket River went from being trashed and underutilized to a space that offers cultural programming, urban agriculture, recreational opportunities, and savvy design to mitigate some of the impacts of the climate crisis. Brownfields are an expensive, challenging opportunity, and these projects demand tremendous patience. But it's an opportunity to center the needs of the people who've had to live with the legacy of industrialization and an extractive economy. It's an opportunity to put resources back into the neighborhood, bring folks together, and develop a collective vision for the future.

DIDI PERSHOUSE

SHE/HER/HERS

"When we allow life underground to thrive, we're positioning our communities to be more resilient"

Didi Pershouse is a healer, **soil sponge** strategist, author, educator, and facilitator. For more than twenty years, she was an acupuncturist and holistic healthcare provider. Her practice, the Clinic at the Center for Sustainable Medicine, was surrounded by organic vegetable and perennial flower gardens on Wabanaki and Abenaki lands in the village of Thetford Center, Vermont. It offered sliding-scale community acupuncture, hands-on healing, and general health, nutritional, and resilience counseling. Over the years, the focus of the clinic expanded from treating individual patients to restoring the health of the social and ecological systems that patients are a part of. In particular, Didi dove deep into exploring the relationships between human health and capitalism, shifting weather patterns, and soil health. Her brilliant book *The Ecology of Care: Medicine, Agriculture, Money, and the Quiet Power of Human and Microbial Communities* explores these connections in great depth.

Didi's work has since shifted to offering training, facilitation, and strategic support to community leaders around the world to improve soil health, and therefore public health, food and water security, climate resilience, and other connected issues. Through the Land and Leadership Initiative, a deeply participatory school that she founded, Didi has developed cutting-edge and accessible curricula, like *Understanding Soil Health and Watershed Function: A Teacher's Manual*, which has been utilized in over sixty countries, and she has traveled around Turtle Island

and beyond to facilitate multistakeholder workshops to figure out how to better support communities of life within and above the land. I had the pleasure of attending one of Didi's online workshops and found myself in a Zoom room with dozens of folks from around the world, all with markedly different relationships with the land but eager to better nourish the health of their patch of soil. I was able to witness Didi's magic in motion; by practicing deep listening and asking really good questions, she is able to find alignment among seemingly disparate people, including farmers, ranchers, policy makers, investors, scientists, and environmentalists across the political spectrum.

Content Notes: Mention of domestic violence

On the water cycle in relation to the changing climate

A big piece of what got me into soil health work was the flooding from Tropical Storm Irene, which hit Vermont in 2011. I had been spending a lot of time thinking about **peak oil** and how a lack of oil supplies would impact the public health system and how we could begin building more resilient healthcare systems, but I hadn't really been thinking much about climate change until that point. I naively assumed that Vermont would maybe be a little warmer, and I thought, *That might be nice*. And then I saw the rivers flooded. There were propane tanks, soccer balls, lawn chairs, and porta-potties floating down the river. It was a bizarre overlay over this typically pristine landscape. 500 miles of roads were washed out, and 250 bridges. It was the moment that it really sunk in that climate change really is going to impact all of us.

Shortly thereafter, I went to a talk by the environmentalist, journalist, and founder of 350.org Bill McKibben, who explained that the symptoms of climate change were going to be felt most intensely in the water cycle. It has since become my work to help illuminate this, and also to show how the broken water cycle is also very much *driving* changes in climate.

On the soil sponge as a climate resilience powerhouse

There is a tremendous amount of life underground that is mostly invisible to us. One of the things that this life underground does is take mineral particles, broken-down rocks, and organizes them into a living matrix that is very much like a sponge; it doesn't fall apart, and it has pores of different size and springiness. It's like taking flour and turning it into bread. They're taking these separate, little, powdery particles and literally gluing and weaving them together into something that is permeable and has incredible structural integrity. So when life is allowed to flourish underground, then when it rains, or floods, or the land is irrigated, or there's morning dew on the grass blades, that water can infiltrate into the soil sponge, and the landscape can store water. It basically creates a reservoir of water underground, which can help protect the land against drought.

For instance, there's a family farmer in so-called North Dakota named Gabe Brown who has become a leader in the soil health movement. He decided to try to better support the life underground, spending years diligently increasing the organic matter in his soil. When he started, his ranch could infiltrate water at a rate of half an inch per hour. Today the ranch's water infiltration rate is eight inches per hour. What that means is that the land can now hold sixteen times more water, or be resilient amid a flood that is sixteen times larger. And any time that water can move down, that means it's not going sideways, so the soil is much less likely to erode away when it does flood. Therefore, having that structural integrity of the soil sponge also protects the capacity to grow food going forward.

When we allow life underground to thrive, we're better positioning our communities to be resilient amid droughts, floods, and widespread food insecurity. We're also helping minimize the threat of wildfire. Right now I have a lot of dead wood in my yard, and I don't want to remove it because I want the organic matter to return to the soil, but it's been so dry that they're just becoming big kindling piles. When there isn't enough water in the land, when a branch falls off the tree, fungal organisms aren't able to do their work, and so decomposition doesn't happen. When there isn't enough water in the land, trees and grasses dry out at the root zone, the fuel load increases, the green season is shortened, and the fire season lengthens. Not being able to retain water in the soil significantly exacerbates the major wildfire stressors.

Lastly, flourishing life underground supports flourishing life aboveground, and plants are one of the most impactful tools that we have in protecting our communities from extreme heat. As plants transpire, which is basically their version of sweating as they do the work of photosynthesis and growing, the air around them cools off. In a transpiring landscape, the air above the plants is 12°C cooler than the air above pavement and 5°C cooler than the air above bare soil.[1] That is profound. If we're considering how to keep our communities cool and moist enough in our changing landscapes until we figure out how to keep fossil fuels in the ground, we know that we can provide oases of coolness by having more transpiration.

The power of soil cannot be underestimated. Two scientists I work with calculated that a five percent increase in transpiration would, in theory, be enough to reverse the current global warming. .

On the power of the ripple effect

For the past few years, I've held a soil sponge gathering. It's a small, dedicated group; we typically get thirty or forty people to come. The first year, I actually put two rather poorly made videos of the conference online. One was called "Restoring Water Cycles to Naturally Cool Climates and Reverse Global Warming" by Walter Jehne. A man in India named Vijay Kumar came across this video online.

Vijay Kumar is a top advisor to the government of Andhra Pradesh, a huge agricultural state in India, and has been instrumental in the natural farming movement in India. Many years ago, he helped to start women's Self-Help Groups (SHGs), which grew to include about twenty million women across the country. Groups of ten to twenty women meet regularly, put a percentage of their income into communal savings accounts, and leverage that savings to get larger loans. When anyone in the group needs resources to get through a hard time, they can tap into those funds. And they have one another to lean on. The program is really successful. As webs of cross-caste relationships have proliferated and strengthened, and as millions of women have become more financially independent, they have taken on issues like domestic violence.

Several years into the program, Vijay learned about zero budget natural farming, a version of regenerative agriculture where farmers were only using inputs that could be made or found within the village, like saved, local seeds, cow dung and urine, forest soils, clay, and other herbal preparations.[2] Vijay started touring little farms that were using zero budget natural farming methods and sending short homemade films to the women's groups. In time, the network of SHGs became the champions for what is now called Community Managed Natural Farming. And soon the state of Andhra Pradesh adopted it as their state form of agriculture. There are now more than 800,000 farmers, or about one in every eight farmers in the state, who are using Community Managed Natural Farming techniques on some level. It's an incredible example of how local support groups can help to achieve broad-scale change.

Vijay watched this talk about restoring water cycles that we posted and decided to try something he calls pre-monsoon dry sowing, or putting seeds in the ground before any rain has come. It worked. And since he was able to tap into the SHG network, there are countless farmers now growing a full crop of food before the rains. This means that they're dramatically extending the green season, growing that much more food, harvesting that much more water through dew capture, cooling the area through transpiration, and growing the soil sponge.

It's just such a reminder that the little things you do can end up having such an impact. Thirty people came to that conference, so it's not about scale. You just need one Vijay Kumar to watch the poorly made video. Forget being perfect, and just do. Share it with the world. Trust that you're in the flow, and focus not on scale or success but on doing the next right thing.

RESILIENCE TOOL SPOTLIGHT

Soil Health and Watershed Restoration

The state of the land has a profound impact on local weather patterns. Climatologist Dr. Millán Millán explains that many of the weather extremes commonly associated with the climate crisis, like drought, flooding, wildfires, and desertification, are not caused simply by greenhouse heating, but by the transformation and degradation of the world's ecosystems. For instance, if a forest is felled and paved over, the land surface heats up, there is less transpiration of moisture into the air, and the warm, barren lands repel clouds, which then retreat and cause precipitation to fall elsewhere. When rain does occasionally fall, the droplets will flow into nearby waterways rather than soak into the ground.

Fortunately, Dr. Millán's research proves that "strategically revegetating even a small expanse of land can make a difference in the surrounding region."[3] There are countless examples that demonstrate the power of soil health and watershed restoration projects in helping reinstate water circulation, bring weather patterns back into balance, dull the impacts of extreme weather events, and support life above and below ground.

Restoration processes vary depending on the land's history, relationships, and functions, but there are a handful of common practices that are generally present. Journalist Judith D. Schwartz, who extensively researched land restoration efforts around the world, gives a handful of simple but profound shared principles across successful projects: (1) find ways to keep water in the ground, (2) build the soil's organic matter (plant and animal material in various stages of decomposition), (3) grow some food, (4) establish trees and other deep-rooted perennial plants, (5) bring grazers onto the land,[4] and (6) increase biodiversity. Together, these practices form a virtuous cycle; grazers help facilitate decomposition, organic matter helps retain water in the soil, plants circulate water from the soil into the air, and a regular flow of water supports life above and below ground.

Didi Pershouse adds two essential principles: (1) practice no till methods of farming and gardening (to not break up the structural integrity of the sponge) and (2) keep soil covered at all times with living plants and plant litter (to feed the biological workforce that builds the sponge). She also encourages folks to spend time deepening their relationship with the land, understanding the land's history and context, and exploring what practices are effective by learning together with others.

Sanjay Rawal, the film director of *Gather*, remarked in an interview with Grist, "North America, Turtle Island, was colonized for its topsoil" that Indigenous peoples had stewarded for millennia.[5] He goes on to explain that it is precisely because colonists rapidly depleted the soil with their farming practices that they expanded further and further west in search of more fertile land, displacing and committing genocide of Indigenous peoples as they went. The state of soil health and watershed function today has been largely shaped by the severing of Indigenous peoples from their traditional lands and ecological knowledge, the privatization of land, the capitalist imperative to extract as much from the earth as quickly as possible, and the rise of mega corporations. Solutions that fail to address these harms are wholly incomplete.

To get involved with soil health and watershed restoration in your community:

- **Research and reflection prompts:** Which watershed are you a part of? What changes have you observed in the water cycle during the time that you've been a part of this watershed? If you can find precipitation tracking for your community online, what trends do you notice? In moments of drought or heavy rains, what have you experienced or observed around you? During extremes in the water cycle, which communities are the most impacted, including more-than-human beings, and how?
- **If you have the passion and access to some land nearby:** Per Didi's recommendation: "Create a restoration management plan for a place nearby. This could involve creating a rain garden to soak up rain off a roof that is puddling nearby, improving infiltration rates or length of green season in a playing field, coming up with a plan to decrease erosion at the edge of a parking lot, or working with a local farmer to improve soil health on a corner of a farm through cover cropping, integrating animal impact, or some other means." For more ideas and guidance, check out her *Understanding Soil Health and Watershed Function* e-book, which is available for free at www.didipershouse.com.

- **If you have a lawn:** Also per Didi's guidance: "There's a new phrase out there, *raise the blade,* which means put your lawn mower setting higher. You can double the amount of leaf area in your lawn by cutting your lawn higher. Or less often. If you can find someone who's making compost and put that compost onto your lawn, then you will be helping repopulate some of the microbial workers and their enzymes, so they can build a more effective sponge under your lawn. You can take some of your lawn or all your lawn out to grow local food, but with the caveat that your garden always needs to have plants growing in it."
- **If you're engaged in a learning community:** Partner with others to organize a larger restoration project at your school or in the neighborhood. If you're a science, history, or health teacher, there are ample opportunities to weave soil health into the curriculum from preschool through college. Again, Didi's *Understanding Soil Health and Watershed Function* e-book contains a myriad of great exercises and lesson plans that you can utilize.
- **If you have access to financial resources:** Contribute or set up a recurring donation to Indigenous-led land restorations taking place in your region. If you can't find a project near you, there are countless projects across Turtle Island and beyond to get involved with, such as the Wolakota Regenerative Buffalo Range and Wildlife Sanctuary, where the Rosebud Economic Development Corporation of the Rosebud Sioux Tribe is working hard to bring back buffalo.[6] For more ideas to support Indigenous-led land stewardship efforts, check out the resilience tool spotlights on *land back* (p. 51) and *good fire* (p. 92).
- **If you have the time and crave camaraderie and guidance:** Check out Ecosystem Restoration Camps at www.ecosystemrestorationcamps.org and find an upcoming gathering near you.
- **General best practices:** Start with your own backyard or city block. Judith D. Schwartz writes, "We've been trained to believe that finding

solutions is a job for experts, but . . . restoration can begin anywhere. Damaged ecosystems can be rejuvenated at all scales, from a small plot between sidewalk and curb to areas large enough to be labeled on a world map. We can all find a garden to tend."[7]

CECI PINEDA

THEY/THEM/THEIRS

"Transformation requires stewardship and regular effort"

Ceci Pineda is a singer songwriter, earth worker, **popular educator**, and cultivator of community at the intersection of QTBIPOC liberation and climate justice. For several years, they worked with BK ROT (and served as the executive director from 2019 to 2021), a youth- and bike-powered food waste collection and composting service based on Lenape land presently known as Brooklyn, New York. BK ROT is responsible for diverting from the landfill over a million pounds of food waste generated by homes and businesses and creating hundreds of thousands of pounds of high-quality compost for local agriculture and soil restoration projects. While the organization is still small relative to the city's vast waste footprint, BK ROT is building a powerful model for a closed-loop, fossil-fuel-free service, and they're creating good jobs for local young folks and connecting more people to the land in the process. I first got to know Ceci through BK ROT's Instagram page, where Ceci espoused not just the tangible ecological benefits of compost but also the myriad lessons they'd gleaned around queerness, ancestry, and interdependence while serving as a steward for the decomposition process. I was moved by their reverence for and deep relationship with the more-than-human world and the soulfulness and heart that they bring to the composting process.

Before joining the BK ROT team, Ceci spent time volunteering at Hattie Carthan Community Garden, a grassroots, BIPOC-led agricultural project in Central Brooklyn, and apprenticing at Soul Fire Farm, an Afro-Indigenous–centered community farm committed to training and supporting the next generation of Black and Brown farmer-activists. Ceci also helped steward Interlocking Roots, a network for QTBIPOC farmers, foodies, and earth workers to connect with one another over food, life stories, and queer ecology and strategize for collective liberation. Ceci is currently focused on releasing an album of songs that explore their connection to change to: Tlalli, meaning "land" in Nahuatl. Their music is a medium for meditation, prayer, and communication with earth kin, and they hope that it will aid others in deeply experiencing their climate-related emotions.

Content Notes: Allusion to anti-queer and anti-trans discrimination

On earth work as medicine

I grew up in places where I had a lot of access to being with the land. When I moved to Brooklyn, I remember that access felt very scarce, so I got involved in a lot of community gardens, and particularly Hattie Carthan Community Garden. I felt like I was in an oasis, a portal. In that space, I was actually able to connect to my gender. In visiting childhood memories, I was able to open up to ancestors in ways that I hadn't before. I connected with the earth and our plant kin. It was a place where I felt like I belonged in my whole self.

Initially I approached the work almost from an egocentric or human-centric place, like *I'm going to heal the earth!* But once I was actually doing the work, it was incredibly healing for *me*. When we do community-based, regenerative earth work, we're doing so much to repair relationships with the land, with each other, with ourselves. In fact, when I first started composting, I would feel this euphoria and joy after hand sifting. I was like, *Wow, this is because I'm connecting to the earth!* Which is true. And there's also bacteria in the soil that literally produces serotonin in our bodies.

On compost as climate action, ancestral connection, and metaphor

There are so many ways in which community composting is climate action and a tool for resilience in the face of increasingly extreme weather. First of all, when we don't compost, our food waste usually ends up in landfills, where it produces methane, which is a very potent greenhouse gas. When we compost, there are some carbon dioxide emissions produced, but only about 5 percent of the greenhouse gas emissions that would be produced in a landfill. That's huge. We also know that finished compost supports the roots of plants in storing more carbon in a long-term form in the soil.[1]

Compost also has incredible water retention capacity (compost can actually hold up to 20 times its weight in water!) so through drought or extreme floods, it supports the land in moving through those circumstances. Compost also helps establish new trees and support trees in having larger canopies. I thought about this a lot while working in New York City, where heat stress has become so consistent that BK ROT had to implement extreme weather protocol for our riders collecting food waste by bike and doing manual labor outdoors. It's now incredibly hot almost every day in the summer, and it's only going to get worse. But shade from trees can help. So there are some very practical ways in which compost itself can help address the impacts of the climate crisis.

But I also think about the deeper things that are happening when people come together around community compost. It was pretty remarkable for me to see this play out in a city context at BK ROT. We specifically work with a lot of Black and Brown youth, and it has been really beautiful to see how they build relationships with earthworms and the land and each other through the process of composting. Composting is about allowing ourselves to see ourselves as part of this older lineage, restoring a connection to something that all our ancestors used to do.

It's also an incredibly accessible system for communities to implement. I almost feel powerless in front of so many of the systems that I desire changing, like the energy grid. But we can choose the way in which we want to relate to our waste. Compost just requires a shovel and a pitchfork.

Working with compost, I feel like a different lesson emerges every day. I think there are so many lessons around interdependence, in particular. For one, as humans, we're simply the stewards of the composting process. Our job is just to create an environment where microorganisms can thrive, where bacteria and fungi can effectively break down the food waste. I also think about how much more efficient community composting is than individual compost systems, as you need a certain amount of input to generate the amount of heat needed. If you're composting alone or in a single-family home, it's going to be a much slower process than if you were combining your food waste with your neighbors'. There are so many ways in which society tells us to live in these individualistic ways, but composting is a consistent reminder of the community ecosystems that we belong to.

Composting also reminds me that transformation takes time. The compost process requires constant attention and work. You can't say, *I'm going to transform this today*, and then be done. We have to revisit it and monitor it and tend to it on a weekly basis. Transformation requires stewardship and regular effort.

On queer community and just climate futures

When I think about QTBIPOC communities, I think about how, for us to even exist, we've had to create our own worlds. To survive and thrive, we have had to build interdependence and community with each other. We have had to literally build our own systems of care and safety, our own identities, our own language. We have had to create our own sanctuaries. I think almost anyone with a queer identity is already creating another world that affirms us, where we belong. That type of imagination is so fundamental in bringing forth just climate futures.

I sometimes think about why I have the identities that I have and am here at this moment in time. I identify as gender nonconforming. In my family, at least on my dad's side, everyone else is lighter skinned than I am. I think there's a part of me and the identities that I hold that are ancestral memory. Because I know that queerness is something that is ancient. There are so many examples in our histories of third genders, and examples in our plant and animal and fungi kin that prove that there are genders beyond the binary, where I find my own affirmations. There's something inherent in our queerness where we both carry this ancestral memory and also this seed of how we can do things differently, to live and love in ways with more integrity, that feel more rooted in the earth.

RESILIENCE TOOL SPOTLIGHT
Community Composting

Compost is a dark, nutrient-rich material produced by the decomposition of organic materials. It's typically made from a combination of nitrogen-rich "green matter," like food scraps, coffee grounds, garden trimmings, or manure, and carbon-rich "brown matter," like wood chips, dried leaves, straw, hay, or cardboard. When water and air are added, decomposers like insects, fungi, and bacteria are able to break down and transform the organic material. Sometimes dubbed "black gold," quality compost is extremely helpful in supporting soil health and plant growth.

Composting is an ancestral practice that land-based cultures have stewarded through history and continue to utilize today. *Community* composting refers to a hyperlocal, participatory, and deeply intentional composting practice and movement that is quickly growing. While community composting efforts manifest in many ways, from bike-powered cooperative businesses to educational programs in school gardens, they're unified by a common ethos. The Institute for Local Self-Reliance, which has become a national hub for community composting efforts, defines the guiding principles of community composting as follows: (1) Food scraps and other organic materials are diverted from landfills; (2) Organic materials are valued as a community asset; (3) Finished compost is used to enhance local soils, support local food production, or support natural ecology; (4) Compost systems are community-scale and are designed to meet the needs of a self-defined community (which might be ten blocks in a dense city or a fifty-mile radius in a rural area); (5) The program should engage and educate the community; and (6) The program should be supported by the community it serves and should serve the community in turn.

Community composting carries the same merits as other forms of composting, plus several more. When the compost process and product are kept as local as possible, communities reap the benefits of improved soil, flourishing plant life, and more bountiful crop yields, all the while becoming less reliant on expensive fertilizers. The land can better retain water and is less likely to erode, therefore diminishing the potential impacts of both droughts and floods. Since food is the single largest category of material placed in municipal landfills, making up an average of 22 percent of total landfill waste,[2] diverting food from the landfill can meaningfully reduce local methane emissions and decrease the number of garbage-hauling trucks on the road.[3] Community composting programs also equip community members with new skills and create good jobs. Many initiatives are also designed to center education, youth involvement, and connection to the land and one another.

To get involved with composting efforts in your community:

- **Research prompts:** Are there any composting programs already in existence or emerging in your area? Do any of the programs appear to be aligned with the community principles outlined by the Institute for Local Self-Reliance? Which organizers are working hard to advance the composting movement in your community? If you can't find any composting projects in your area, might policy-related barriers be a factor?
- **Journal prompts:** As Ceci points out, there are countless lessons to be gleaned from compost. What have you previously viewed as disposable that you'd like to reframe and reclaim as an asset? The practice of composting is a sacred ritual of decomposition from which new life emerges; what systems, paradigms, or mindsets would you like to see decay and decompose to make way for the new? How might you help nourish this process?
- **If you're interested in learning more:** Take a training course that will equip you with the skills and confidence to begin composting at home, like a local composter certification course or an online course like the Institute for Local Self-Reliance's Community Composting 101. Or get involved in a community composting project! Ceci says, "Seek out neighbors to work on things together. Find spaces where this work is already underway and where you might be comfortable getting engaged. Don't get discouraged; if one space isn't right for you, there are others that will hold you. There are so many people who desire this connection and who are nourishing this connection. And the earth is everywhere around us. The land is always calling us back. It's about listening to that calling and also being a little courageous to try new things."

- **If you're passionate about composting and you have the bandwidth:** Help start a community composting program in your neighborhood. If you're involved with a school, community garden, or nearby farm, that can be a great place to start. Check to see if there are any community compost grant programs available in your municipality or state. Consider how existing rules and regulations might affect your operation, which sites might work best, who will be involved and how you'll engage your community, which benefits you'll be maximizing for (education, soil improvement for local farms, waste diversion, local job creation, etc.), where you'll be sourcing composting materials, where finished compost will go, and how it will all be transported. Explore detailed how-to guides from the Institute for Local Self-Reliance, James McSweeney, and Robin Greenfield. If you're considering starting a new organization, check out the webinar *Entity Structure for Community Composters* cohosted by the Institute for Local Self-Reliance and the Sustainable Economies Law Center. Try to connect with folks who have engaged in community compost efforts elsewhere. Start small.
- **If you have access to land:** Begin a compost system at home. Since composting is most efficient with more material, talk to your neighbors about combining materials to produce compost together.
- **If there are policy-related barriers to composting in your community:** Become a composting advocate. Contact your local planners and elected officials and attend town meetings to call for funding, technical assistance, training, and long-term land access for community composting programs, as well as the amendment of any unnecessary policies and regulations that make it difficult for community composting programs to operate.

MARGO ROBBINS

SHE/HER/HERS

"Burning is an important fire prevention solution—we also burn to connect to the land and our original agreement to care for each other"

Margo's passion for good fire (fire put on the ground intentionally, generally as sacred ritual and care for the land and water) is positively infectious. It's nearly impossible to converse with Margo or listen to her speak and not feel compelled to witness the medicine of good fire for yourself. Today's wildfires can be a tremendous source of fear, sorrow, and very real loss, but Margo is unwavering in her commitment to show that fire can also be a source of regeneration, reclamation, connection, and outright glee. She is the cofounder and executive director of the Cultural Fire Management Council, an organization based on the upper Yurok Reservation with the mission of facilitating cultural burning on the Yurok Reservation and ancestral lands, ultimately restoring Yurok ancestral territory to a healthy, viable ecosystem that supports the cultural lifeways of Yurok people. CFMC works hard to train community members in cultural burn practices, strengthen state and federal support of cultural burning, transfer knowledge across generations, and communicate the importance of good fire to the public. Margo is also the cofounder and co-lead of the Indigenous Peoples Burn Network, a support network led by Native American people who are revitalizing their traditional fire cultures in a contemporary context with the goal of assisting Indigenous nations across the so-called United States and abroad to reclaim their traditional fire regimes.

When Margo is not putting fire on the ground and spreading her good fire wisdom, she is the Indian education director for the Klamath-Trinity Joint Unified School District, a basket weaver and regalia maker, and a mother and grandma.

Content Notes: Settler colonization, anti-Indigenous racism, and violence

On her relationship with fire

I grew up in the traditional village of Morek on the Yurok Reservation along the Klamath River, in far northern California. Now I live right up the hill from my homeplace. Our landscape has changed dramatically in recent years because of the suppression and exclusion of fire. The non-Native people who came here did not understand that fire is meant to be part of the ecosystem, that as humans we're also meant to be part of that system, and that we all must work together to keep it physically and spiritually in balance. They were afraid of fire, so they went to war on fire. I was just rereading a letter written by the Forest Service in 1918, the year my mother was born. The letter instructed his people out in the field to shoot Indians that were seen starting fires. Later the instruction changed from shooting Indians on sight to imprisoning Indians seen starting a fire. I don't know if it's worse to be killed or imprisoned, but neither one of them is good. That was an effective way to stop the use of fire by Native people, who were taking care of the land and keeping it healthy, productive, and in balance.

That's the practice that has been in place for over 100 years. Now we live in this landscape where the brush is so thick that you can't see or walk through it. The animals can't live there. All the deer have gone someplace else to live. We used to use elk for food and blankets, but we don't have any elk on the reservation anymore because they like prairie land. Almost half of our land used to be prairie, and now there are just slivers of prairie here and there. Our prairies are only 3 percent of what they once were. Fire suppression and exclusion have done a lot of damage, not only to the landscape but also to the animals who no longer have a hospitable place to live. As Native people, we depend upon the plants and the animals for our cultural lifeways, so it has really been devastating on so many levels.

I am a basket weaver, and the materials that we use to make baskets are fire dependent. I learned how to make baskets as a young girl in school during a period in the 1960s. During that time, someone from CalFire was actually coming up into our area and helping the elders to burn their hazel patches. While he was doing that, we had enough fire to reproduce basket weaving materials. We had hazel sticks for the basket frames and soft and pliable bear grass to weave with. After he retired, that was the end of that.

When fire was removed from our ecosystem, our ability to carry on the tradition of basket weaving grew slimmer and slimmer until it was at risk of getting lost. I had always intended to learn how to make baby baskets but when I learned that I was going to become a grandma, we didn't have the materials. I was so thankful to have sticks that had been given to me from when my grandmother passed on, but it felt really wrong that we couldn't even make our traditional baskets because we're not allowed to burn.

At the same time, we were also getting really worried about the state of the landscape. With all the brush building up, and only one road through the reservation, we worried that our elders might not be able to escape if there was a wildfire. Those were the two driving forces behind our community's decision to make it our number one priority to bring fire back to the land. As a basket weaver myself, I was driven to make it happen, no matter the obstacles. We weren't going to take no for an answer.

Today we're working to reclaim our traditional fire practices. For the past ten years or so we've held Cultural Burn Training Exchange (TREX) sessions in the fall and the spring. We use these trainings for qualified firefighters as an opportunity to restore the land and increase the health and availability of culturally important species. We'll pick places to burn that are abundant in cultural resources like hazel, wild iris, mushrooms, and medicine plants. Fire allows us to better access the food and medicine growing around us. We also like to burn prairie land in the hopes that the elk will be able to come home.

Fire reconnects us to the land and reminds us of our original agreement to care for each other. The health of our people is connected to the health of the land and fire is going to make our land healthy again. Fire is going to reconnect people to the spirits of the land.

Burning is also an important fire prevention solution. We need to shout that out as loud as we can and as often as we can. When we burn, we're burning the understory. It clears out brush and dead, woody stuff, but the trees are still intact. It's fire prevention through fuel reduction. It decreases the likelihood of fire reaching up into the treetops. These major wildfires are different; they're traveling through the tops of the trees and they're burning everything.

On the importance of intergenerational leadership

Fire is incredible on so many levels. One piece of that is its accessibility. Everyone can be a part of fire in any way that they choose. At one of our recent burns, my daughter had her baby in her backpack, I brought my three-year-old grandson, and my sister, who's older than me, was out there, too. We had three generations of people learning about fire together.

It's really important that our work be intergenerational. We realize that the path that we've set for ourselves to restore the ancestral territory of the Yurok, which is about half a million acres, isn't going to get done in our lifetime. So it's crucial that we pass along knowledge and experience to the next generations.

RESILIENCE TOOL SPOTLIGHT
Good Fire

For thousands of years, cultural fire and prescribed burns, also known as "good fire," have been implemented strategically by Indigenous peoples to maintain healthy, productive, and balanced ecosystems. The low- and moderate-intensity fire clears fuel to minimize the danger of high-intensity fires, allows medicine plants and traditional food sources to thrive, purifies waterways, and supports the lives of countless more-than-human beings. Cultural fire is a sacred practice. For the past century or so, however, the US federal fire policy has hinged upon a strategy of fire suppression, designed in part to protect commercial timber supplies. The fire suppression regime has been violently enforced, and cultural fire and prescribed burns have been restricted.

Today Indigenous peoples still face persecution and penalty for traditional burning practices, even upon reservations. But as wildfires rage through overstocked forests, the need for prescribed burns is becoming more widely understood. In 2012, after five years of research, the Jemez Fire and Humans in Resilience Ecosystems project concluded that "living without smoke and fire is simply not an option," particularly as temperatures increase and rainfall becomes more infrequent and extreme.[1] Continuing to safely reside in fire-dependent ecosystems will require folks to embrace prescribed fire as a necessary, beneficial medicine and a disaster-prevention strategy.

Restoring balance in fire-dependent ecosystems also means respecting and centering the Indigenous science and forest stewardship that prevented dangerous wildfires for millennia. Fortunately, fire advocates like Margo are working hard to reclaim Indigenous practices, shift policy, and create training programs so that more folks can safely get involved in putting low-intensity, preventative, ecosystem-balancing fire on the ground.

The reintroduction of good fire is not without its challenges. As the climate crisis worsens, it becomes harder and harder to safely conduct controlled burns. Additionally, as humans move further into wildland areas, it becomes more challenging to burn at the scale necessary to effectively reduce the risk of high-intensity fire. However, as Bill Tripp, the director of the Karuk Tribe Department of Natural Resources, says, "We're not going to be able to solve this problem overnight. This is going to be a 100- to 200-year process. But we have to start."[2] And many of us will need to pitch in.

To get involved with good fire efforts in your community:

- **Research prompts:** Who are the original stewards of your watershed, and what did their relationship to fire look like prior to settler colonization? What does their relationship to fire look like today? How do federal and state policies and budget allocations support or restrict fire suppression versus prescribed fire and fire-prevention efforts in your region today? Which Indigenous-led organizations, advocates, and practitioners are leading efforts to restore good fire in your region today?

- **Journal prompt:** What have you been taught about the relationship between ecosystem health and human presence? How might you reframe or reclaim narratives about human and larger ecosystem interdependence with what you now understand to be true?
- **If you have access to time:** Contact your local, state, and federal representatives via email, letter, or phone to call for Indigenous fire sovereignty and to ask that resources be invested in an expansion of prescribed burns and prescribed fire training. Attend town hall meetings or office hours to ask what your representative is doing to support good fire efforts.
- **If you have access to land:** Participate in a Prescribed Fire Training Exchange workshop to learn how to safely put fire on the ground in your yard or larger community.
- **If you have access to financial resources:** Support the work of organizations like the Indigenous Peoples Burn Network that are conducting cultural and prescribed burns, advocating for law reform to allow Indigenous fire sovereignty, and leading prescribed fire trainings or otherwise educating folks about good fire.

Economic Regeneration

In order to evade environmental collapse, we must move away from economic systems that are powered by fossil fuels, tethered to an imperative of endless growth, and designed to amass tremendous wealth and power for the few at the expense of the many, and we must move toward economic systems that are just, participatory, decentralized, and designed to create the conditions for all sources and forms of life to thrive. These essays are from entrepreneurs, strategists, and organizers who are resisting so-called climate solutions that fit neatly within the extractive economy and who are architecting futures in which economic, ecological, and social well-being can coexist and flourish.

DORIA ROBINSON

SHE/HER/HERS

"We needed to create the capacity to make change ourselves"

Doria Robinson, a third-generation resident of Richmond, California, is committed to sowing the seeds of a Just Transition in her hometown. As the executive director of Urban Tilth for the past fifteen years, Doria has nurtured and expanded its ecosystem of food sovereignty, climate justice, and community power–building programs. She's grown and strengthened their training and apprenticeship offerings, created dozens of good, local jobs, and developed community-supported agriculture (CSA) and farm stand programs to feed hundreds of local families and support a constellation of small, nearby farms. She has helped transform disinvested spaces across Richmond into flourishing gardens, farms, and vibrant social hubs. In particular, she helped deeply engage the community around the restoration process of the Richmond Greenway, formerly a railroad line and now a three-mile biking and walking trail flanked by thirty-two acres of edible and flower gardens, recreational spaces, and community-designed artwork. When Doria and I first spoke, she was lovingly tending to the summer's strawberry crop in one of the Greenway's edible gardens known as Berryland, a 2,000-square-foot berry oasis.

Doria is also helping create pathways toward a regenerative economy and community-controlled wealth as the cofounder of Cooperation Richmond, an organization dedicated to supporting local residents in building cooperative enterprises. She's working to lift up frontline community-led solutions and Just Transition strategy as a steering committee member with the Richmond Our Power Coalition. She is a fixture of the Richmond community, and as a member of the Climate Justice Alliance and other networks, Doria is also highly engaged in the cultivation of movement strategy across Turtle Island and beyond.

Content Notes: Structural racism and classism, mention of drug epidemics and the school-to-prison pipeline, death of loved ones

On the wisdom instilled in her by her elders

I grew up a few blocks from this garden. This used to be a broken-down railroad track that I used as a shortcut to my grandparents' church, also a few blocks from here. My grandfather designed that church even though he wasn't an architect, and we built it ourselves. He and the other elders were very resourceful. Their mantra was, "It doesn't matter if you don't have a lot. Together we can do things."

In fact, when they moved here from Louisiana, they came with a bunch of families and pooled their money to buy properties together. They were not wealthy people, or even middle class. They came from places which had only recently gotten indoor bathrooms. But together they were able to buy a 350-acre ranch in Fairfield. That's where I spent all my time growing up. They had chickens and turkeys and horses and cows who were all free-range. All their food was grown organically. They used horse manure and food scraps instead of fertilizer and were making their own compost. It was more out of frugality than anything else.

Between those two worlds, growing up at the ranch and also in Richmond in the '80s, with the crack epidemic, and drive-by shootings, and the backdrop of the Chevron refinery, it created the person I am.[1]

On the real impacts of false climate solutions

We are definitely feeling the impacts of misguided climate solutions in Richmond. One of the biggest things that we are feeling and will continue to feel is California's ridiculous cap-and-trade program.[2] We are one of the disposable communities that gets to receive all the extra pollution that people are trading. They are trading our health. The Chevron refinery, which is the biggest point source of greenhouse gas emissions in the state of California, gets to pollute extra here because they are supposedly reducing emissions somewhere else.

Unfortunately, these false solutions are going to continue. They're going to continue because there's so little audible pushback to it in the mainstream media.

People blindly accept this idea without thinking about who might be impacted. It's like people don't even care what happens to the frontline communities.

They say that infectious diseases will also increase because of climate changes.[3] With the current Covid pandemic, we're already seeing how this will impact communities like ours, which already have increased exposure to chronic illnesses. We have one of the highest rates of asthma, and so our community has been extremely affected by the pandemic.[4] My aunt, who had very bad, chronic asthma from growing up here, just passed away. She was diagnosed with Covid and, two or three days later, was gone. My uncle has been in intensive care for almost a week now on a ventilator. If we didn't have to fight Chevron around having our health matter, if we didn't have our hospitals taken away, if we didn't have such poor access to healthcare, maybe they wouldn't have been in such bad shape.

I'm so mad at Chevron. They just made a whole Black Lives Matter statement. Nothing they've done for the last hundred years has shown that any of the Black lives that live right here matter. They are one of the worst polluters of all the refineries along this whole corridor. They are a multibillion-dollar, multinational company, and they just drip out pieces of money to shut people up. If they really cared, they would update their hundred-year-old facility so it made the least impact on our health. If our Black lives really mattered, Chevron would take steps to actually protect our lives.

On Just Transition solutions, trans-local organizing, and the cooperative movement

Several years ago, Movement Generation, an organization that Urban Tilth has a relationship with, began to work really intentionally with other grassroots groups to define the Just Transition. (Note that Movement Generation Justice & Ecology Project is a strategy-shaping, alliance-building, and educational hub for movement leaders, popular educators, and organizations committed to the liberation and restoration of land, labor, and culture.) A group of us started doing these informational exchange visits. We called it trans-local organizing. We visited urban farms in Detroit. Separately, with the Chorus Foundation frontline communities cohort, we connected with folks in Kentucky who are in the coal mining area and Native Alaskan communities who are impacted by the petroleum and steel industries. These exchanges planted so many ideas in our head.

Kentucky, in particular, was an eye-opening visit. Our communities are supposed to hate each other, but when we got there, we saw that it was a lot like Richmond, except the people there are mostly white. They have a school-to-prison pipeline, high unemployment, issues with drug abuse. So many things are similar. It was pretty profound to be able to see that and then try to figure out how to build together and be real allies for each other.

In those exchanges, the beginning of the Just Transition framework started to emerge—organizing in a way where we're providing for people's needs in what we're doing, not just fighting for someone else to change the rules. We needed to create the capacity to make change ourselves. I think it opened up this idea that we really had to be unafraid to engage directly with the economy. How can we provide our own power and water and food? The way that our larger society is engaging with those things now is entirely based around profit-making, which drives extreme extraction. There's no room for the environment in that model; it's not a factor except as a resource. So to really have a say in all of that, we realized that we needed to engage. We needed to create more projects and programs that were actually solutions. Instead of complaining and saying what we don't want, we also need to develop what we do want. And it feels so much more empowering when we can say *I can go do something about this* rather than waiting for the next election.

Inspired by this ethos, I founded Cooperation Richmond. I was reflecting upon the failures of the cooperative movement to really meet the needs of low-income people of color to start cooperative businesses. They don't have access to capital. They don't have access to bank loans.[5] They often don't even have access to those city-run, low-income bank loan programs because their credit is bad. They can have a dream all they want, but they'll never be in charge of their own labor unless some wealthier folks let them in, which also doesn't happen. Because they often don't present with the same kind of educational background or the same communication style, they don't land those entrepreneurship competitions.

When we started thinking about the things that you really need to put in place for low-income people of color to successfully start a cooperative business, we began to dream up a program that really hinged upon mentorship. We didn't just want to design a kick start or bootcamp, which are great for getting an idea out, but we wanted folks to be connected with people who had traveled the path before, who could give advice, so that when things inevitably got complicated or overwhelming, they could navigate that together. A good mentor would have connections to folks with resources, good accountants, lawyers who know how to set up cooperatives. Those things are important so that people don't have to make every road themself. We created Cooperation Richmond to be just that, focused on coaching and connecting people to capital.

RESILIENCE TOOL SPOTLIGHT
Cooperatives

Cooperative businesses have long been an effective tool for resistance and resilience. In the late eighteenth century, free and enslaved Black folks began pooling resources to grow, sell, and distribute food, purchase land, care for the sick, organize burials, and more. As time went on, cooperatives became a strategy not just for survival and community care but also to organize and train activists, build collective wealth, and divest from white-supremacist capitalist economies. By 1907 W. E. B. Du Bois held a conference on cooperatives and was able to list 154 Black-owned cooperatives in existence, from grocery stores, collective farms, and gas stations to credit unions, buying clubs, housing cooperatives, and more. Du Bois emphasized the importance of cooperative economics in creating prosperity for the masses, broad ownership, economic self-determination, and Black liberation.[6] By 1940 most Black colleges in the United States taught cooperative economic theory in their core curriculum. In the 1960s, civil rights activist Fannie Lou Hamer carried the torch of Black cooperative excellence and founded the Freedom Farm Cooperative to give impoverished Sunflower County residents the opportunity to steward their own land. As gender and racial justice organizer Nia Evans says, "Collective governance is in our ancestry."[7]

Today there is an abundance of cooperative business structures, including worker cooperatives, producer cooperatives, service cooperatives, housing cooperatives, multistakeholder or hybrid cooperatives, and beyond. They are generally bound by the same set of seven principles originally derived from the Rochdale Society of Equitable Pioneers and now recognized by the International Cooperative Alliance: (1) open and voluntary membership, (2) democratic member control, (3) members' economic participation, (4) autonomy and independence, (5) education, training, and information, (6) cooperation among cooperatives, and (7) concern for community.

Cooperative businesses can play an important role in building just, inclusive, and participatory economies, as well as meeting the essential needs of communities through climate chaos. By their very nature, cooperatives are highly communicative, deeply rooted in community and relationships, and beholden to more than just short-term profits. When emergencies occur or business is otherwise disrupted, cooperatives can act nimbly, prioritize the needs of members (including the health and safety of workers), buffer themselves from volatile markets, and lean on their robust relationship network.

To get involved with the cooperative commonwealth:

- **Research prompt:** Which cooperatives currently exist in your community? How are they structured, what are their espoused values, and how do they embody those values?
- **Journal prompts:** Reflect on how cooperative principles like shared ownership, democratic decision-making processes, cooperation, and concern for the larger community show up, or are absent, from the communities (which could include family, friends, and neighbors) and organizations that you're a part of. What might it look and feel like for those communities and organizations to embody cooperative principles more fully?
- **If you're excited to support cooperative businesses:** Make a list of the goods and services that you regularly buy, and look for cooperative businesses in your region that you could frequent (and potentially

become a member of) instead. Commit to buying all your holiday presents from cooperative businesses and see how it goes. If you purchase goods or services for your place of work, research how you might shift more company dollars toward values-aligned, cooperative businesses.

- **If you're keen to uplift cooperative values:** Start a conversation with coworkers or fellow organizers and activists about what it might look like to begin applying more cooperative principles in your organization. Particularly if you're in a position of power in your organization, explore what it could look like to formally transition to a cooperative structure. Sustainable Economies Law Center's Cooperatives Program offers an abundance of resources to get started down this path.
- **If you have access to financial resources:** Consider investing in or gifting money to organizations like the Seed Commons that are helping to fund and actualize cooperative businesses around the country.

CASSIA HERRON

SHE/HER/HERS

"To holistically address climate change, it's clear that we need to fix our democracy"

I was instantly smitten with Cassia Herron after learning that her strategy for racial justice, community wealth creation, and climate action follows in the footsteps of civil rights activist and solidarity economics champion Fannie Lou Hamer; Cassia is all about cooperative systems, Black liberation, and access to good food. As a community organizer, urban planner, writer, educator, entrepreneur, and mother of two, Cassia has spent decades working to shift the food landscape of Louisville, Kentucky. In 2016 she cofounded the Louisville Association for Community Economics (LACE), a nonprofit cooperative developer with a mission to educate the community about cooperative economics and support Lousville's growing cooperative ecosystem. In 2019 Cassia and her comrades incorporated the Louisville Community Grocery and assembled an eight-member board of directors to bring to life a full-service, community-owned grocery store in an area of the city that grocers began abandoning in the 1950s as the city became redlined, leaving predominantly Black and working-class neighborhoods with few nourishing food options.

While Cassia is deeply rooted in West Louisville, she's also building people power throughout the state and the US South. As the former chairperson of Kentuckians For The Commonwealth, she has been instrumental in strategizing a Just Transition away from the coal economy, centering the needs of Appalachian workers and families while fiercely advocating for environmental and climate justice. Since Cassia was selected as one of eight recipients of the Common Future Bridge Fellowship, she has also been building relationships, dreaming, and creating with other local economy leaders across the South to develop blueprints for more equitable, inclusive communities.

Content Notes: Structural classism, mention of racialized murder by police

On her journey into just-agriculture and energy-transition movements

I was born and raised in Richmond, Kentucky, and have been in Louisville since 1997, when I came here to attend the University of Louisville. Toward the end of my undergraduate program, I met an organizer with the Community Farm Alliance who was regularly coming to meetings held at my church, Quinn Chapel AME Church, where I was working and had been worshipping since coming to Louisville. I joined the CFA team, and we began organizing community residents and farmers to create the first farmers market in the Portland neighborhood. We eventually expanded to other neighborhoods and found that the markets were a wonderful way to bring community leaders across the city together, to start envisioning what they could do in cooperation with one another. We saw the potential of farmers markets to reinvigorate commercial corridors and stimulate economic activity. And we saw these markets as an additional tool to help Kentucky farmers transition from a tobacco-based agriculture economy to one that's more food-and-fiber based.[1]

Years later, after a decade in the agriculture space, I watched as our legislature decided that they would not comply with the Clean Power Plan and they wouldn't be submitting a plan to transition Kentucky's energy economy.[2] Meanwhile, despite the growing awareness around the dangers of pollution and climate change, they weren't taking any action other than increasingly deregulating polluters without any acknowledgment of the damage that was causing. I thought it was all pretty asinine. I had already been part of one major economic transition in Kentucky and saw what was possible when we were proactive. We made sure that the massive tobacco settlement funds were invested in our agricultural transition to keep our state's small network of 85,000 farms alive.[3] I knew that if we didn't begin proactively transitioning Kentucky's energy economy, the climate crisis would change things for us.

So I began organizing with Kentuckians For The Commonwealth (KFTC), which has been working on environmental and climate justice in our state for decades, particularly around the issue of mountaintop removal in Appalachia. As I got involved with the KFTC committee working on an Empower Kentucky Plan, which laid out a just energy transition for the state, I became so impressed by the amount of work that regular, ordinary Kentuckians had been doing for decades to investigate the coal economy's actual impacts on our state, to transition Appalachia, and to put forth a vision reflecting what folks wanted to see happen. I eventually joined the KFTC leadership as the vice chair and then was elected board chair, and my work has since expanded to engaging with Kentuckians to better understand our collective responsibility to participate in the transition of our economies, whether that's the energy or agriculture or service economy.

We're also working hard to improve our democracy and the way it works for us. We continue to send the same old white men to represent us, and we have yet to see a shift in priorities. Part of KFTC's work is making sure that we have good candidates running for office and that people understand the candidate options that we have. To holistically address climate change, it's clear that we need to fix our democracy.

On the importance of cooperative economics and decentralized food systems

While organizing with the Community Farm Alliance, we worked hard to ensure that farmers could diversify their agriculture businesses, but I noticed that there was work to be done to ensure that Kentuckians have access to food. Over the last two decades, I've seen communities across the state lose their grocery stores. People have to leave their county boundaries to shop for groceries at a Walmart or shop for groceries at the Dollar General store. People have less and less control over the types of food available to them, and the situation has only gotten worse in recent years.

So in 2017 I cofounded the Louisville Association for Community Economics (LACE), both to educate our community about cooperative economics and to use cooperative economics to transform the community's economy, beginning with the community's first cooperative grocery. The Louisville Cooperative Grocery will be a full-service grocery store that sells healthy, local, and affordable food to residents of a neighborhood that experiences high rates of food insecurity. There will be an amazing deli so that people can sit down for lunch and then pick up dinner on the way home.

Cooperative ownership is not a new concept; it's centuries old. But it is relatively new for cooperatives to exist in a super capitalist society that has made us all believe that if we just work hard enough, we'll be rich and able to provide for our families. What I've come to know and understand is that we have an unbalanced distribution of resources, and the cooperative model is one way to try to even that out at least a little bit. I don't even have a car, but because I've been able to partner with other people in my community, I'm an owner in two businesses.

I've also come to understand that the more that we're able to decentralize the food system, the better access we'll have and the more control we'll have over the food that we eat. The food system has been centralized so much that every item in your grocery bag can be owned by the same parent company. The centralization of the food system means that we can have thousands of cattle ranches nearby, but when we buy steak at the store, we won't know where it's from, how it was raised, or how far it's traveled. It means that we're completely removed from the impacts of growing and transporting mangoes and avocados all year round. It means that there used to be upwards of ten thousand different varieties of corn and tomatoes, but we can only find the same few varieties in the grocery store. That loss of culturally significant food alone makes it clear what our role must be in decentralizing and recreating the food system. So when we build locally owned grocery stores or farmers markets or community-supported agriculture programs, it may feel like a small step, but together we're building a movement that's diversifying and decentralizing our food system so that we have more control over what it looks like.

On the power of local coalition-building

Particularly in a place like Kentucky, the greatest source of power comes from poor people organizing together. There are "progressive" groups, Black groups, Brown groups, LGBTQ+ groups, and more, but the power resides in us being a collective unit. When we create multistakeholder coalitions of people who share the same visions and are affected by the same things, that's how we build power.

For example, after Breonna Taylor was killed, my sister Keturah Herron led the campaign to limit the use of no-knock warrants in our community. After two weeks of campaigning, the Louisville Metro Council was set to vote on Breonna's

Law. Hundreds of us were gathered on the steps of the mayor's office, watching the meeting play out inside the chambers. We watched all twenty-six council members unanimously approve the ban on no-knock warrants. And then, since it was also budget season, we watched as activists testified about how they wanted to see the city's money spent. In a matter of days, the council allocated a $3.5 million bond for a community grocery. We still don't know for sure whether we'll receive that money, but we know that allocation wouldn't have been made if it weren't for the fact that people were organizing around Breonna's Law, showing up to the meeting, and tuning in to watch the ordinance pass in the name of justice for Breonna. It's all about the power of the people.

MARTA CERONI, PHD

SHE/HER/HERS

"If I live in an interconnected way, most of the things that I do create a ripple effect"

Dr. Marta Ceroni serves as the co-director of the Academy for Systems Change, where she coordinates research and communicates the systems thinking work of the late environmental scientist and author Donella Meadows. Marta also brings together cohorts of systems leaders to support and build connective tissue across movements. While working under Marta's guidance as a fellow for the Academy for Systems Change during my final years of undergraduate school, I was struck by Marta's ability to lead with tremendous humility, warmth, and humor and to embody her values with joy and pleasure.

Prior to joining the Academy for Systems Change, Marta studied and taught about economies that prioritize ecosystem and community well-being. She consulted with the United Nations Development Programme and spent a decade as a research professor at the Gund Institute for Environment at the University of Vermont. After the State of Vermont adopted the genuine progress indicator framework in 2012,[1] Marta was instrumental in developing the tools for Vermont to apply the framework to public policy and budget analysis. Throughout her career, Marta has immersed herself in local and regional organizing and building at the intersections of energy democracy, food sovereignty, and cooperative economies.

On fossil fuel resistance and dancing with systems

I grew up in Milan, Italy, and also in the countryside of Cremona and Parma. I tend to trace back my interest in climate action to an injustice that I suffered when I was a young girl in the outskirts of Milan. It was a very polluted place. A nearby oil refinery released fumes every single night, and our home did not have good windows. We would have to breathe in these disgusting fumes. It felt so unjust that something you breathe, and that you cannot escape from, could be forced upon you. The outrage came at an early age.

I currently live in the Upper Valley between New Hampshire and Vermont. Decades after that experience, I find myself organizing against the presence of the fossil fuel industry in my community once again. In fact, just ten days ago we celebrated a victory against liquefied natural gas pipelines that were going to be built underground nearby (Liberty Utilities had planned to build the twenty-seven-mile "Granite Bridge" pipeline along Route 101). It took five years of organizing. At some point there were eight different groups working on this. We ultimately coalesced thanks to a wonderful organizer who helped us come together and get to a beautiful place of victory.

I had a friend recently ask me if I am in a fight or if I am in a dance. I am a dancer; I dance with systems. The purpose of a dance is to create a shared experience. We had to dance to defeat the pipeline project because we didn't have a unified strategy. We had eight groups, so we had to accept duplication and the lack of a spokesperson. We had to get comfortable with the fluidity and the beautiful mess. We had to trust the wisdom of the group, or the many groups.

On the scale of change

I used to wonder how to balance the local work with the impact that I'd like to see at a global level. And then I started questioning the whole thing because it made me feel powerless. I think I've given up on that aspiration, actually.

I've also begun questioning the whole concept of individual or personal impact. It feels pretentious to be in the individualistic mindset that I could even own a certain type of impact. I've started to build my humility muscles, because in reality, this is work that can take generations and has taken generations.

I've begun to think that if I live in an interconnected way, most of the things that I do create a ripple effect around me, and I don't have to feel responsible for creating all the impact. Therefore, a big question for me right now is how enterprises can start to mimic the web of life, so to speak.

One-on-one relationships are crucial, but I think that we really create that ripple effect when we build multistakeholder networks. The co-op model is particularly suited for that, in part because it's built out of cooperative principles, but also because co-ops can create a web of relationships that support a thriving local economy, people, and community in times of need. For example, the food co-op supports farmers, employees, and members. That model creates resilience for so many different community members, and in this present moment, as we deal with the climate crisis, Covid-19, and an economic recession, resilience has become such an important, practical necessity.

On systems thinking and its limitations

There are different angles to systems thinking. One piece is the diagnosis; being able to really understand what is happening requires the ability to step back and see the systems. Donella Meadows's work is an incredible source for that wisdom. The questions that she raises are foundational to being able to identify the underlying structures that drive the behaviors that cause the conditions to persist, despite the best intentions of most or many people. There's a lot of investment in the current status quo, so it's essential to unpack those investments and the structures that preserve the status quo. Without that lens, we risk implementing quick fixes. People are fed up with quick fixes. That's why we talk so much about systems change and radical transformation.

That's not enough, though. You can be analytical all you want, but unless you really explore the human dimension of systems, you can bypass the players in the system, their lived experience, and their traumas. You don't have to map a whole system to understand it, but you do need to experience it through the lenses that are important. When you work in community, you start to understand what's at play and what's at risk. That's why it's super important to combine the systems lens with a deep understanding of the humans involved in the system.

And yet! Even that is not enough! Donella Meadows missed the dimension of unpacking her own privilege. Although she experienced poverty and she experienced marginalization, she never really brought that perspective back to her work. That's something I'm discovering in her work right now.

In fact, we're in the midst of a deep moment of reflection and shaking things up as we reckon with our lineage. Our founders came from an MIT-educated, Western science background. If systems approaches hold the promise to shift things, then it is absolutely necessary that we apply a deep equity lens. Right now we're looking closely not only at systems thinking tools, but also at all the additional layers that allow us to show up for systems change effectively: deep listening, compassionate leadership, effective communication, and connecting to the wisdoms that are not typically manifested in more conventional or Western ways of problem solving or systems intervention. It's been a disruptive time for the Academy for Systems Change, one that holds a lot of promise.

RESILIENCE TOOL SPOTLIGHT

Fossil Fuel Resistance

To avoid the most extreme climate scenarios and preserve the livability of the planet, it is imperative that the global use of fossil fuels swiftly and steeply declines. Specifically, in order to have a greater than 50 percent chance of limiting warming to 1.5 degrees Celsius,[2] the global use of coal, oil, and gas must decline about 95 percent, 60 percent, and 45 percent, respectively, by 2050 as compared to 2019 levels.[3] To meet these targets outlined by the Sixth Assessment Report of the Intergovernmental Panel on Climate Change, it will require stranding most fossil fuel infrastructure and leaving considerable amounts of coal, oil, and gas in the ground that fossil fuel corporations have already purchased.

Fossil fuel projects are responsible for more than just massive amounts of greenhouse gas emissions. They spew toxic pollution into the air, water, and land, harming public health and destroying the habitat of more-than-human community members. They also often desecrate sacred Indigenous sites, violate treaties with tribal nations, and threaten traditional ways of life. In the so-called United States and around the world, fossil fuel extraction and production disproportionately harm low-income and BIPOC communities.

Unfortunately, and hardly surprisingly, the fossil fuel industry is adamantly resisting the much-needed transition away from fossil fuels. Fossil fuel corporations are still working hard to grow their markets for oil and gas around the world. They are actively expanding the infrastructure that drives new fossil fuel extraction, like pipelines and terminals, while negligibly investing in renewable energy.[4] In the so-called United States, state and federal governments are complicit in the expansion of the fossil fuel industry. Not only do they continue to approve thousands of oil and gas drilling permits and new infrastructure projects each year, but the United States government spends billions of dollars directly subsidizing the fossil fuel industry each year.[5]

At the same time, communities on the front lines have been organizing against the fossil fuel industry for decades, and the grassroots movement is growing. They have been fighting to close coal plants, halt construction of new oil pipelines, and stop the expansion of refineries. They have been working in tandem with labor unions to build a Just Transition, or pathways to phase out extractive, exploitative, and polluting industries while creating good, living-wage jobs for their workers. Recognizing that they are up against an industry with tremendous political influence and financial resources, the fossil fuel resistance has embraced a variety of tactics. From #NoDAPL to #StopLine3 to #SaveTheArctic and beyond, activists have rallied in the streets of state and federal capitals, camped by construction sites for months at a time, chained themselves to construction equipment, locked themselves inside corporate offices, launched legal battles, flooded public hearings, pressured institutions to divest from fossil fuel companies, and held prayerful ceremonies.

And they have won. In fact, a 2021 report by Indigenous Environmental Network and Oil Change International found that, had the fossil fuel projects stopped or delayed by Indigenous-led fossil fuel resistance been actualized, they would be emitting additional greenhouse gas emissions equivalent to about 28 percent of annual United States and Canadian emissions, or about 400 coal-fired power plants.[6] While state and federal policy, international agreements, and investor influence have thus far failed to meaningfully rein in or shift the direction of the fossil fuel industry, grassroots organizing has worked.

To get involved with the fossil fuel resistance movement:

- **Research prompts:** Are there any campaigns against new or existing fossil fuel projects currently underway in your region? Which groups and/or activists are leading the fight? What are their main concerns? What kind of involvement and support are they calling for?
- **If there are fossil fuel resistance campaigns underway in your region:** Follow the lead of organizers to determine where and how your help may be needed. There are typically many ways to lend support, both from afar and on the ground, from showing up and engaging in direct actions to cooking for the community, making artwork for posters and social media, providing logistical and administrative support, offering pro bono legal support to activists facing charges, and so much more. Before joining (or organizing) a protest or other in-person uprising, consider gathering comrades and brushing up on community care and strategy skills together via Training for Change's online toolkits and workshops.
- **If you have the financial resources:** Send money and supplies to the front lines. Fund the work of Tribal Nations, Indigenous water protectors, land defenders, pipeline fighters, workers' rights groups, local environmental justice organizations, and other grassroots movements working to "keep it in the ground" and build a Just Transition.
- **If you (or your money) are affiliated with large institutions:** Take a moment to jot down the institutions that you have some stake in that may or may not have money invested in fossil fuel companies. Did you attend a college with a large endowment? Has your workplace enrolled you in a pension plan? Are you a member of any large museums? Do you belong to a faith-based organization? Where are you keeping or investing your money? Which credit card do you use? Then, search www.gofossilfree.org to find the divestment status of the institutions you've noted. Find more info about mutual funds, managers, and ETFs at www.fossilfreefunds.org. Next, check to see if any active divestment

campaigns regarding these institutions are underway. If so, ask how you can join or lend support. Last, make your voice heard. Send an email, write a letter, tweet at the organization, or start a petition. And if the institution refuses to divest (and it's feasible for you), take your business elsewhere. Just be sure to let them know why you're leaving.

- **If you're active on social media:** Hold fossil fuel companies accountable for their **greenwashing** by responding to their posts on Twitter or your social media platform of choice. Climate writer and activist Mary Heglar calls this practice "greentrolling"[7] and explains that responding directly to companies' real-time communications is particularly effective for a few reasons: it exposes more people to their corporate hypocrisy and helps erode public trust in Big Oil; social media allows for a two-way conversation, sometimes even forcing companies to respond to viral tweets; and the phenomenon attracts newcomers to the climate movement.[8] Plus it's pretty fun and easy. While you're at it, consider engaging with politicians who are taking donations or otherwise profiting from the fossil fuel industry while actively opposing climate policy.
- **If you have a few minutes:** Advocate to keep fossil fuel dollars out of government. Ask your elected representatives and candidates running for office to sign the No Fossil Fuel Money pledge and refuse campaign contributions from oil, gas, and coal industry executives, lobbyists, and PACs.

CRYSTAL HUANG

SHE/HER/HERS

"The DNA of energy democracy is about people having power, literally, financially, and politically"

Crystal Huang's enthusiasm for energy democracy is, dare I say, electric. I was stoked to talk with Crystal because she's thinking about and organizing around energy in a much more holistic and radical way than many folks in the renewable energy space, by which I mean that she's skillfully scheming to address the roots of dysfunction in energy systems rather than simply working to swap oil and coal with solar and wind. She has long been a champion of renewable energy, with over a decade of experience in climate solutions technology. Today, however, she is less focused on the technology itself and is more interested in models of energy ownership that can serve as a pathway toward community self-determination, abundance, stability, and well-being. Crystal, a fervent believer in the power of the people, views the climate crisis as a pivotal opportunity to center equity and liberation in the redesigning of society, and she advocates for democratic energy ownership as a central pillar of a more just future.

At the local level, Crystal is helping build a model for grassroots-led, community-governed, and fossil-fuel-free energy as the cofounder and president of People Power Solar Cooperative, an organization based on Ohlone land in so-called Oakland, California, in neighborhoods disproportionately impacted by the harm of the fossil fuel industry. At the national level, she is working to connect grassroots efforts and advance the larger energy democracy movement as the national coordinator of the Energy Democracy Project, a dynamic collaboration between about forty organizations engaged in energy justice work.

Content Notes: Structural racism and classism, mentions of violence, mass incarceration, and the immigration crisis

On her journey to cofounding People Power Solar Cooperative

Most of my career has been in solar and renewable energy. The work that I did used to be very global, very high level, primarily focused on sales and marketing strategy analysis. I actually used to work at the largest solar manufacturer in the world. Then I spent some time with a startup incubation around solar and utility integration, and meanwhile I was helping produce a movie around climate solutions called *Time to Choose*. The intention of the movie, made by Oscar-winning filmmaker Charles Ferguson, was really to do something similar to what Al Gore had done with *An Inconvenient Truth*, which woke a lot of people up about climate change. We wanted to wake a lot of people up about climate *solutions*, about how we already have the technology that's needed. We wanted to get people to understand that they could save more money, that it would be better for their health, and it would be better for the planet to be carbon-free in their mobility, food, electricity, heat, and how they reinvest their money. It was going to be a very compelling documentary focused on individual behavior change.

Up until that point, the work that I was doing was very global, so there was a lot I didn't get, particularly about the current system that we all rely on. The theory that our film team came in with was really that, sure, low-income communities cannot afford electric vehicles and solar yet, but they can start changing their diet to be plant-based and stop eating McDonald's. As if that's a choice. These statements of course reflect a lot of ignorance.

One of my jobs was to do local activation. But when I started connecting with local community organizations and showing up at their meetings, I saw community organizations like Phat Beets Produce in North Oakland and Alemany Farm in San Francisco literally fighting neighborhood violence by creating community gardens and growing food. That completely blew my mind because I realized it was everything that we wanted these communities to do, except that they were coming at it from an angle that mattered to them, not from our angle, which was more about climate solutions and personal health.

When I saw that our solutions to different problems were the same, it made me realize that climate change is not the problem. Climate change is the symptom of the very system that's also causing neighborhood violence, mass incarceration, violence against women, the immigration crisis, displacement within our cities, literally every single issue. They're all connected to the same things. I knew that I had a lot left to learn, but it was clear that the problem was not a lack of understanding of what's happening. The problem was not about visibility. The problem was power and privilege. And the solution was to put power back in the hands of the community. So I left the film project and started to get myself deeper and deeper into the organizing world, really connecting with folks and strategizing around what building community power means.

On (truly) community-owned solar as a transformative climate resilience solution

What we are trying to do with People Power Solar Cooperative is very different from clean energy access. It's about energy democracy. And ultimately, the DNA of energy democracy is about people having power, literally, financially, and politically.

When we talk about community-owned energy, we must be really specific about what ownership means, because people can own stock but may not have very much power at all over that company. Our definition of community ownership is that the decision-making, voting, and governance is democratic. Also, for people to really own energy, they must understand it. When we're talking about something as amorphous as energy, it's really important for people to connect to it in a different way than we're used to. Our culture has trained our brains to believe that energy is just something we should care about when the power goes out. Most people don't understand how vulnerable we are without it. It's more than just charging our phones and electronics. It's how we access and preserve food, how we move around, in the clothes that we wear. Everything in our homes is related to energy. We take that for granted. We don't understand what energy really means, and that is what makes it really disempowering.

When people get to own the energy and have community ownership of energy in a truly cooperative ownership form, their relationship to energy changes. As they steward projects themselves, they get to understand the nitty-gritty of what energy means and are completely aware of how vital this resource is. And when they own it, they get to make collective decisions about how they want the benefits to flow. For instance, many of our member owners are currently pushing to build a community-owned refrigeration system in someone's backyard to finally have a place for local Black and Brown farmers to bring and store their fresh produce. When people build a real relationship with energy, they dream beyond just solar panels and getting financial benefits back. They can literally design something that works for their specific community. That's how you can create community resilience.

Traditionally if you wanted to go solar, you had to own a property to put it on. Our model basically allows anyone to be able to steward an energy project, even if they're not a property owner. So renters or even unhoused people can connect with property owners in their network, which could be relatives or a house of worship or a community center or friends, and say, "We can put energy on your roof, reduce your electricity bill, and at the same time, you could participate in a way to build community power." In some ways, we're literally disconnecting the ownership of land to the ownership of power. A lot of what we're doing as a cooperative is allowing people to understand that we should not have to wait for corporations and nonprofits to come save us. We can actually solve this issue ourselves, collectively.

That being said, there are a lot of barriers that we're working against. There's a power struggle from corporations, and it's still really difficult for communities to have control over something like energy in the state of California. We don't even have a viable community solar law here. On top of that, the mainstream narrative is that you must make money from energy, that energy is a commodity that you sell. Energy is so vital to our ability to survive. It's a huge contradiction. And as long as people tie the concept of energy to money, it's really hard to break out of that to see that energy can build wealth far beyond financial wealth.

On ancestral knowledge and people power

One of the reasons that I find so much love and hope in the communities disproportionately oppressed by the system is that they still have connection to ancestral wisdom. They still have the knowledge that is deeply connected to how we used to be before we unconsciously decided to destroy the world. As long as those seeds are still there, there are people who will keep showing up with a true understanding of what people power means. People power is not just a million people signing a petition. People power is not going out on the street holding a sign. It's people understanding what we deserve and that we can actually take it back by collectively building wealth with each other.

RESILIENCE TOOL SPOTLIGHT

Community-Owned Renewable Energy

Energy systems are at a critical turning point. Amid an urgent need to transition away from fossil fuels and toward renewable energy sources, there also lies an opportunity to rethink how our energy systems are structured, both physically and procedurally. Advocates for a sustainable and just energy future are fighting not just to replace coal, oil, and gas with renewable energy sources like solar and wind but are also calling for the literal and figurative decentralization of power predicated upon the understanding that reliable energy access is a basic human need and right.

Over the past 120 years, the energy industry of the so-called United States has evolved to be largely investor owned, highly consolidated, and politically powerful. At the turn of the twentieth century, utility companies were clamoring to bring electricity to every building in every city across the country,[1] and so they looked to wealthy investors to help finance the expansion of electricity systems, which were composed of massive coal and hydroelectric power plants, high-voltage lines that could transmit power across long distances, and smaller lines that would distribute energy to individual buildings. By the 1930s, just ten central utility companies controlled 75 percent of the country's energy utilities.[2] Not surprisingly, the overcentralized, investor-owned utilities (IOUs) began price-gouging their customers. They also inserted themselves into energy policy and regulation at the local, state, and federal levels to maximize profit for their investors.

While utility companies continue to unduly influence policy-making today to preserve their status quo, the movement to return energy ownership, benefits, and decision-making to the people is growing. Many folks are tapping into the power of longstanding energy models, like public utilities and cooperatives, that lend themselves much better to public accountability than IOUs. Community choice aggregates (CCAs), which are now authorized to exist in ten states, are also becoming a popular tool for communities to exercise sovereignty over how their energy is generated and to develop renewable and **distributed energy** resources locally. Finally, community renewable energy programs, which are authorized in about a third of the states, allow folks to buy or lease a portion of a renewable energy project and then receive credit for the energy provided to the grid.

These alternative energy models are imperfect, and many are utilized in ways that are just as extractive in practice as traditional utility companies and energy projects, but these structures have the potential to be a force for energy democracy and myriad other benefits. When energy projects are locally generated and community governed, they help weaken the political and economic power of energy monopolies while bolstering the political and economic power of the people. Locally owned renewable energy projects are more likely to create jobs for community members, use local materials, borrow from local banks, and circulate profits within the community. And while fossil fuel markets are erratic, renewable energy costs tend to be stable, if not steadily decreasing, over time, which increases energy accessibility and reduces the **energy burden** on low-income and BIPOC communities. When people develop energy projects in their own communities, they also have tremendous incentive and built-in accountability to make every effort to minimize possible harm to the people and places they hold dear.

To get involved with the energy democracy movement in your community:

- **Research prompts:** (1) Which entity are you currently receiving energy from? How is the utility structured? How are they generating power? If it's an IOU, check out websites www.followthemoney.org and www.littlesis.org to better understand your utility's influence on politicians and policy. If it's a rural electric cooperative, research the board members to better understand their values and priorities; find out when the next board member election is and who's running. (2) Does your state authorize community choice aggregates (CCAs) and/or community solar programs? Are there any CCAs or community solar programs in your region? The Institute for Self-Reliance's Community Power Map is a great resource to better understand the policy landscape! (3) Are there any other emerging community-owned energy projects near you?
- **If your community is particularly prone to unreliable energy or power shutoffs:** Consider teaming up with friends or neighbors to start a battery sharing collective. Portable, solar-powered batteries can be handy, if not lifesaving, amid an energy emergency, but they can also be expensive. Backup battery collectives allow you to spread both the cost and benefits among community members. Check out Shareable's guide, "How to Start a Back-Up Battery Collective," available at Shareable.net, to get started.
- **If you have the passion and bandwidth:** Organize to shift policy and leadership in energy decision-making spaces at the local and state levels. Rally around energy democracy champions to get them elected onto local commissions, boards, and legislatures (or consider running yourself!). Organizations like Solar United Neighbors, the Energy Democracy Project, and Sustainable Economies Law Center have an abundance of resources to get started in the fight to make community-owned renewable energy projects truly viable.

- **If you have access to the financial resources:** Redistribute funds or invest in community-owned energy projects advancing renewable energy and energy democracy in your region. Use the Institute for Self-Reliance's Community Power Map, available online, to find projects near you.
- **General best practices:** Aviva Chomsky writes in her 2022 book *Is Science Enough?*, "Because of the fundamental role of energy in our economy, energy democracy relates deeply to the struggle for social, racial, and economic justice."[3] Make sure to center those most impacted by social, racial, and economic injustices in your community's struggle for energy democracy. Check out the People's Utility Justice Playbook, available as a PDF at www.energydemocracy.us, for more best practices and ideas to combat the tactics of investor-owned utilities.

MOJI IGUN

SHE/HER/HERS

"Everything that we do is connected to this problem of how we manage and utilize our resources wisely"

Moji Igun is a zero waste, sustainability, and circularity consultant, small business owner, international speaker, and educator. I originally connected with Moji through social media, where she was uplifting the importance of waste minimization in the context of larger systems rather than fixating on individual behavior changes. It was a delight to see Moji expanding the conversation beyond mason jars and plastic straws and toward ancestral practice, capitalism, circular design, imagination, and the lessons that she's learned as the founder of Blue Daisi Consulting, a firm she created to support organizations in finding creative ways to keep trash out of the landfill and move toward a closed-loop business design. Moji specializes in one-on-one coaching and small business strategy but also offers community workshops and free online classes.

Moji also serves on the board of directors for Zero Waste Washington, where she helps shape and drive strong zero-waste policy. When she's not working hard to shift the systems that make it difficult for people and organizations to eliminate their waste, Moji finds inspiration and energy in community with Black women and creative people. She also prioritizes plentiful time to rest, dream, and connect with the mountains and water surrounding the Duwamish lands of so-called Seattle, Washington, where she resides.

On small business ownership and creating your dream role

I learned about the whole zero-waste craze around 2017. I saw people on social media fitting all their trash into a mason jar. I was enchanted by that idea because it made waste so visual and tangible. I could identify the problem. I could see when I was doing better. It worked well with my brain.

I became a little obsessed, and I started asking, *Why is this so hard for me to do on an individual level? Why isn't this just the way that things are?* I realized I wanted to explore these questions more deeply. At first, I started interviewing my friends, local businesses, my favorite poke bowl shop, and blogging about it. I just wanted to better understand how it all worked. But then I became so engrossed in the problem that I wanted to figure out how to solve it. I couldn't find any jobs working on this, and the closest things didn't feel like the right culture fit for me. So I reverse-engineered the job I wanted. I realized that, through my background in K–12 education, engineering, and commercial construction, I had the perfect skill set to help people, and specifically small business owners, work through these problems.

For folks who are considering creating their own environmental dream job, I would give two pieces of advice. First of all, know that it doesn't need to be a *job*. I wouldn't always recommend entrepreneurship. It's not at all the easiest path. So consider what it might look like to create your dream *community role*. You can volunteer. You can create art. There are endless possibilities.

If the idea of creating your own position really resonates with you, though, I highly recommend tapping into a process of rest and imagination. Use those tools to daydream things that don't exist. We do need roles that don't quite exist yet or not in the forms that make sense for the way our world works. After a whole lot of brainstorming, work backward to figure out how you could possibly get there, and then try it.

On reclaiming the zero-waste movement

When I first learned the term *zero waste* and came to understand it as a specific concept, one of the reasons that I gravitated toward it so much is because it's the way I was raised. My parents would pack my lunch in ziplock bags, and I would have to bring all that stuff back home. I'd wash them and dry them and use those plastic baggies over and over again. That is zero-waste behavior. We used our plastic bags from the grocery store to line all our waste bins. Being resourceful is just part of the way I was raised. It's inherently part of a lot of Black, Brown, and Indigenous cultures. It's a way that so many people already live. So when I saw a name attached to the behavior, I was like, oh, we can call it zero waste! That's wonderful!

The term has become divisive, though. I talk to so many people in this space, including leaders that I look up to in the zero-waste movement, who have just abandoned the term or idea altogether because they're tired of having the same unproductive conversations over and over again. People still come to me and say, "I can't fit my trash in a jar, I'll never be like you." I've never said that I can fit my trash in a jar! I don't aspire to that! And people are surprised by that. They'll respond, "What do you mean? Aren't you the zero-waste girl?" I am, but not like that. The work that I'm doing is with businesses and governments. I'm working to shape city policy. So when I say zero waste, I'm talking about building a society that facilitates zero waste. Not *you're a bad person for having some waste*, but *let's make it so easy for you to be zero waste so that it's not even a question, it's just how things are*. For me, it's about moving past guilt and shame so that we can be more productive in our conversations. Nobody wins an award for fitting their trash in a mason jar, so let's stop fixating on whether we're bad people because we have trash in our waste bins. Instead, let's talk about why it's so hard to significantly reduce or eliminate our waste, and what structures are in place that we can change.

I also find myself regularly clarifying that waste is so much more than just an issue of the **Great Pacific Garbage Patch**. That makes for great marketing, but it doesn't land close to home for most people. It makes sustainability feel like an "over there" problem instead of a "right here" problem. When we talk about waste as a faraway problem, it becomes less tangible. In reality, I can connect waste to just about anything that someone could care about. Waste is central to everything because it's grounded in the way that we consume. And everyone needs to consume. Everything that we do is connected to this problem of how we manage and utilize our resources wisely. So if you don't care about the sea turtles, that's fair. But perhaps you care about why landfills are located where they're located, or air pollution from incineration, or the extraction of resources themselves. Maybe it's a money thing, or a cultural thing, or an ease thing, or an aesthetic thing. I care about making waste relevant to what you care about and helping you connect those dots. At the end of the day, if the motivation is different for different people, but we're getting the result that we want, I don't think that's a bad thing.

On prioritizing slowness and ease

I can't take any credit for my rest practice. The ideas all come from the Nap Ministry led by Tricia Hersey. When I need to remind myself to practice those ideas, I go to the Nap Ministry account and scroll or pull out my copy of *Rest is Resistance*.[1] I remind myself that I feel so much more centered when I make time to rest. For me, that looks like making time to take walks in the middle of the day, or listening to my body and taking a nap when I need one, or journaling in the morning to gather my thoughts, or saying no to projects that are going to push me over the edge.

Tricia Hersey frames rest as pushing back against capitalism. For me, it also means creating the world that I want to live in. I don't want to live in a world where I feel rushed and tense at all times. I want to live in a world with ease. My rest practice helps me create that world on the smallest scale possible.

Slowing down also shifts our relationship with waste. Using myself as an example, when I have a big project and I'm really stressed or I have to stay up late, those are the times when I order a bunch of junk food wrapped in packaging. Or I forget about the stuff in my fridge and it all goes bad. When I'm rushed, those are the moments when I get careless and can't move with intention. I think that's common among a lot of people. Our world is not designed to allow us to slow down, so we have to find ways to force ourselves to pause.

Collective Care

We got us. For generations, communities have been developing innovative ways to care for, nourish, and protect one another when the state fails to do so, or is the source of violence itself. These essays are from farmers, healers, and educators who are carrying that legacy forward. As their communities deal with escalating climate calamities and compounding crises, they are building webs of connection and support, preserving and lifting up cultural wisdom, and carving out spaces for joy, softness, and sanctuary.

DESEREE FONTENOT

SHE/THEY

"Monocultures of the land and mind leave us the most vulnerable to crisis"

Deseree Fontenot is a weaver of liberatory movements. Their knack for cross-pollinating and collaborating across typically siloed spaces emerged as a student organizer in college as they rallied the queer student assembly to engage in the anti-sweatshop movement and partnered with groups across campus to advocate for gender-inclusive bathrooms. While Deseree came to appreciate the importance of building alliances and dissolving the siloes that can dictate who advocates for what, they were also experiencing the power of sacred community spaces, where folks could embody their whole, authentic selves and experience belonging. Complexity, connection, and deep care have naturally come to be central and defining characteristics of Deseree's work and how they relate in the world.

Today Deseree is a farmer, grassroots ecologist, and organizer. She is the cofounder of the Queer Ecojustice Project, an organizing and storytelling project at the intersections of queer and trans liberation and just ecological futures. She is also an organizer and board president of Shelterwood Collective, a land-based project committed to ecosystem restoration and community healing at the nexus of Black liberation and Indigenous sovereignty. As a collective member of Movement Generation, Deseree cultivates relationships with climate justice organizations across Turtle Island, teaches profoundly inspiring and perspective-shifting ecological and political education workshops and retreats (a few of which I've had the immense pleasure of attending), and dreams and schemes up Just Transition strategies.

Content Notes: AIDS crisis, anti-queer and anti-trans discrimination, structural racism

On queer ecology as affirmation and inspiration

Queer as a noun, for me and coming from many different voices and ancestors, is a medley of gender and sexually diverse ways of being: expressions, cultures, bodies, intimate and erotic interactions, and relationships. It can encompass many identities, behaviors, and processes. As a verb, *to queer* is to destabilize, it's to disrupt, expand, bend, blend, and transform what has been normalized and naturalized. And the definition of *ecology* that I typically use in my work is the study of the relationships of home and the knowledge of home. So with queer ecologies brought together, the definition that hits home for me the most is a lens through which we can reclaim and reimagine our bodies, lands, communities, and movements with the knowledge that diversity assures resilience. And when I talk about diversity, I hope to get out of the ways in which that word has become a trigger point for a lot of us, and instead talk about the inextricably linked biological and cultural diversity that is the basis of our survival on this planet.

When I started farming, I saw that nature was the queerest thing on earth. Fungi have 36,000 forms of embodiment and sexual, asexual, and nonsexual strategies. Grizzly bears will raise young in communities of up to six other mothers, generally not as sexual partners but as co-parents. And avocado trees will open up their pollen-producing flowers by day and will open up their pollen-receiving, fruit-producing flowers by night, shifting forms of sex expression within a twenty-four-hour period. When we study ecology, there are countless types of embodiment, of kinship, of ways that we care for each other, of bonding, affection, and courtship, that hold lessons for us about adapting, surviving, and cooperating on this planet. It's the breadth of this diversity that ensures resilience and balanced ecological systems.

So why, when we get to human diversity, does it become a crime against nature? This narrative stems from a worldview that is centered on monocultures: the idea that there's one model of development that's based on endless growth and consumption, the belief that one assimilated world culture is ideal, the practice of planting the same crops across huge swaths of land, etc. Monocultures of the land and the mind have the same effect on our planet: they leave us the most vulnerable to crisis. Biological and cultural diversity are inextricably linked in a way that makes it impossible for us to believe that one way of being, thinking, or doing is ideal for any system.

On the intersections of queer liberation and climate justice

I spend a lot of time talking specifically about how queer and trans folks of color, in particular, are on the front lines of the climate crisis, because for so long our movements have been super siloed between environmental justice and queer liberation, even though it's always been intertwined. None of this is new. Folks, and particularly Black and Brown queer and trans folks, have been weaving these connections together for a long time.

The ways that the climate crisis exacerbates existing inequities and traumas are true across many communities, and it's no different for queer and trans communities. A lot of queer and trans youth are houseless already, and housing and food insecurity is compounded by extreme weather events and disasters.[1] We've seen over and over again, through winter storms and fires and the pandemic, how incredibly underresourced and neglected folks who are not housed inside are through disasters. Meanwhile, policing of folks on the street is often heightened amid and after crises,[2] and we're already one of the most overpoliced populations in the country.[3] There's also an inadequate emergency shelter system, and many crisis centers are run by more conservative groups. During acute shocks, folks often lose access to hormones and medications, and there can be very real health ramifications. Queer elders and disabled folks also regularly lose access to essential resources, like electricity, during and after crisis. And on and on.

And as queer and trans folks of color, we have so much magic in terms of adapting and creating our own communities of care. This is our jam. Looking back at the height of the AIDS crisis, amid deep stigmatization and ostracization, this radical, queer way of organizing was born. It was not only about making demands for the resources that we need to survive, but also in making it happen for ourselves—practicing sanctuary and building it for each other. With the climate crisis, we need more and more sanctuary spaces, and to me that looks like queer and trans folks, and particularly QTBIPOC, creating land projects in both rural and urban spaces. It looks like mutual aid work. I also think about all the online queer dance parties I attended this past year, where people from all over could tune in and pick up the vibes. We don't always hold that stuff up because it's the fun part, but joy has to be there, otherwise what are we doing? We have to keep it irresistible.

When we think of climate change, we get a lot of bombardment with the doom and gloom, with this idea that there is no future. Likewise, for a good era of queer scholarship and queer studies, this idea of *no futures* was a big theme. The book *Cruising Utopia* by José Esteban-Muñoz was a big challenge to that.[4] It asked, What are we as queer folks bringing into the future, and what do we have as queer folks specifically to offer in a future where adaptation and survival and new ways of being will be central to getting to that world? That's something I continually come back to in my work: How might we broadcast what a queer ecological future could feel and taste and look like?

On the importance of land in justice movements

In 2016 I started organizing around cooperative ownership models and housing with a group called the People of Color Sustainable Housing Network in the Bay Area. We worked on getting more BIPOC folks into some of the existing **limited-equity cooperative housing** that exists in the area as a starting point, and we also had meetups for folks who were interested in building new models of housing, co-ownership, and other interventions to the current housing system, where housing is either incredibly expensive, so cramped that you're living on top of one another, or you're priced out altogether. As BIPOC folks, we have been and continue to be locked out of the system of owning our own nuclear home. Ancestrally, though, we've had a much more communal relationship to land. So it feels important to continually de-silo and think about land justice not just in terms of housing but also in terms of rematriation, and reparations, and ecological restoration, and food sovereignty.

On a practical level, to weather the storms ahead, we have to be prepared to govern, and land is essential to what it looks like to practice collective self-governance. This is nothing new. The civil rights movement could not have happened without Black farmers. I'm constantly being reminded of this. Black farms were spaces where people could take refuge, to hide from folks who were committing violence. Over time, as we contest for power and for the collective right to create the world that we need, it's going to come with resistance, and we will actually need spaces for safety. These land-based projects are not just about having places to drink tea and have fun.

And on a more personal level, I feel like the sense of agency one has is so different if you feel rooted in a place on a daily basis. Your collective well-being becomes tied up with the collective well-being of that place. The more that I put my hands back in the soil and am continually present with the cycles and processes of the land, water, and air, the more energy I have to share the things I share and do the things I do. If I'm not in some constant reflective relationship with place, it's all just ideas. I don't feel attached or grounded in anything. But when I can take half an hour each day to observe, walk, check in on my seeds, and be in relationship to my ancestry, that's where I find my power. That has to be the real fuel and the real wealth of the world.

I always come back to the stories that my grandmother tells about my great-grandparents, Lonias and Azenor Poullard, and the amazing, biodiverse, polyculture gardens that they cultivated on their small plots of land as sharecroppers. Those stories of resistance, resilience, and holding their land-based knowledge and ancestry, even through generations of land-based trauma, inspire me to keep doing this work.

RESILIENCE TOOL SPOTLIGHT

Community-Controlled Housing

Private homeownership is still touted as a cornerstone of the so-called American dream and a primary pathway to financial wealth and security in the United States, but discriminatory public policy, violent displacement, and institutionalized biases have structurally excluded Black people, Indigenous people, and people of color as well as low-income, disabled, queer and trans, and immigrant communities from homeownership for centuries. Secure, long-term housing has only become increasingly inaccessible for many folks who have been priced out or pushed out of their communities. Meanwhile, rising sea levels, extreme weather, and more frequent and intense climate-related catastrophes are expected to increase pressure on land and housing options, particularly for those who have been most structurally excluded.

As affordable housing becomes more elusive for many, alternative housing strategies designed to build collective wealth and resist displacement are emerging. Shared-ownership and democratically governed housing are not new concepts, but modern-day models, like community land trusts and permanent real estate cooperatives, are being tailored to address some of the key challenges of this moment in time: to take land off the speculative market, to decommodify land and bring it into community stewardship, to divest from Wall Street and reinvest in community, to disrupt some of the root causes of racialized inequality, and to buffer community members from gentrification and the climate crisis.

The alternative housing options currently taking root offer tremendous range in size, structure, and setup. For instance, community land trusts are set up as nonprofits and generally provide affordable housing specifically for low- to moderate-income families, while housing cooperatives have more financing options, including borrowing money from community members, and can provide housing to all folks along the income spectrum. The Champlain Housing Trust in northwest Vermont houses over two thousand residents in a combination of affordable rental apartments, co-op homes, and shared-appreciation single-family homes and condominiums, while the Canopy in Boston is a ten-bedroom cooperative specifically for folks seeking a built-in network of support while raising children. This diversity is important; the more types of inclusive, community-controlled options that exist to meet differing desires and constraints, the better positioned we'll be to keep all our neighbors safe, housed, and thriving through the crises ahead.

To engage with the democratically governed housing movement:

- **Research prompts:** What are some of the largest structural barriers to safe and desirable housing in your community? Who is being left out or pushed out? Which people or organizations in your region are advocating for housing as a human right and dreaming up and actualizing more inclusive, community-controlled housing options for structurally excluded folks?
- **Journal prompts:** Take a moment to acknowledge and let go of your preconceived notions about what housing should look like. Then consider: What might housing look like if it were optimized to keep communities safe amid extreme weather events? To ensure ease of life and care for elders, disabled folks, and young people? To foster connection and minimize loneliness? To guarantee healthy, desirable homes for all? How do these housing scenarios differ from what you currently see in your neighborhood?
- **If you have access to time and a passion for housing:** Consider teaming up with similarly passionate and committed community members to plant the seeds of a shared-ownership or democratically governed housing organization in your area. Begin by having as many conversations as possible with community members about what their housing needs and desires look like. Dream big but start small.
- **If you have access to time and shareable skills:** Lend your talents to emerging or existing community-controlled housing projects, which can oftentimes benefit from assistance with legal matters, construction and design, media and communications, community engagement, fundraising, real estate purchasing, and more.
- **If you have access to financial resources:** Consider making a financial commitment to a local, democratically governed housing community, whether by scheduling a recurring donation, paying membership dues to become a community owner, or purchasing ownership shares to become an investor owner.

- **General best practices:** Whether advocating for a greater spectrum of housing options, supporting an existing housing community, or helping sprout a new project, make sure to center the folks who are most impacted by housing exclusion and inaccessibility every step of the way.

JACQUELINE THANH

SHE/HER/HERS

"Environmental justice is reproductive justice is racial justice"

Jacqueline Thanh, also called Jacquie, is a clinically trained human rights advocate who has spent her career championing comprehensive health equity by working intimately with communities affected by the intersecting issues of sexual assault and domestic violence, human and labor trafficking, incarceration, homelessness, and more. She was born and raised in San Francisco, between Little Saigon and Chinatown, cut her teeth in social work in Chicago, where she attended graduate school, and has lived in New Orleans for nearly a decade. While Jacquie, a self-described grief worker, has held space for immense heaviness over the years, she stays rooted and inspired by her ancestors and explains that her greatest source of power is her ability to retain joy in a multitude of hard places. Indeed, her laugh is infectious and exuberance for life palpable. Speaking with Jacquie, I was struck by the need for more grief workers in the climate movement, along with the importance of making space for folks to show up in their full humanity.

Jacquie currently serves as the executive director of VAYLA New Orleans, an organization that emerged during the reconstruction period after Hurricane Katrina as community members rallied to resist a toxic landfill that was supposed to be located less than a mile away from the heart of Village de L'Est, a predominantly Vietnamese American community.[1] Today VAYLA is an incubator for Asian American and Pacific Islander (AAPI) leaders in New Orleans and beyond who are addressing environmental justice, reproductive justice, food sovereignty, and civic engagement issues affecting their communities. Jacquie, who was the first in her family to attend college, is also presently working toward her doctorate on Asian American femme leadership and environmental storytelling.

Content Notes: Racial segregation and structural racism, displacement, food apartheid

On the looming threat of becoming refugees again

I am the daughter of Chinese Vietnamese refugees. Every storm season here in New Orleans, I think about how our communities are at risk of being made refugees again. In 2020 eight storms made landfall in Louisiana. One made landfall in New Orleans, and I remember being in the eye of the storm in my house. This month it has rained every day except for one. As I look out the window, I can see a moat of water forming around our homes.[2] Our coasts are eroding, our barrier islands are sinking, our roads are flooding. It's a real threat to our way of life. Our communities continue to face retraumatization and displacement.

Meanwhile, segregation is quite literally baked into the soil here. I live in a traditionally Black neighborhood, in Gentilly, where people have lived for generations, in homes passed down from the grandmothers of their grandmothers. VAYLA is homed nearby in an area where Vietnamese Catholics were placed decades ago by Catholic charities and the archdiocese.[3] The church was very intentional about where they placed refugees, in parts of the city that had already gone through white flight.[4] The racialization of land use and planning in New Orleans is complicated. And every time there is a disaster, as some folks move out of these areas that are habitually underfunded and under-cared-for, and some folks remain or return because it's the only place where they can afford to live, it just becomes more complicated.

On food as a gateway

One of the most gratifying things to have emerged from the pandemic has been VAYLA's new Farm to Families Initiative. When I was flying back from my wedding back in February of 2020, wearing my N95 mask and wiping everything down, I couldn't stop thinking about all the news centered around China. It was very real to me that some racist stuff was going to happen. I got back and insisted that we put out a survey to our community for hate crimes. Somebody reported that they went to a food bank and were turned away. Another reported that they were served rotten food. It was clear that there were some real barriers developing for our community around safe access to food.

At the same time, there are many Vietnamese farmers in New Orleans East, largely elders who don't have children to pass their farms on to.[5] A lot of the restaurants in the city buy from these smaller farms, but when the pandemic began, that stopped. VAYLA decided to partner with local Vietnamese farmers both to support the folks who grow our food and to deliver mutual aid boxes to our community members. The boxes began with culturally integrative produce. After all, you may not be able to get bok choy at Whole Foods, but you can get it from one of the aunties growing it down the street! The Farm to Families deliveries evolved to include in-language Covid-19 information—there has been a lot of misinformation, loss, and grief in our community—voter registration, and hurricane preparedness materials. Food is the gateway. Food is an opportunity to heal, to begin conversations, to build civic engagement, and to raise awareness around issues important to our community members.

It's important to note that food access issues, as well as creative food access *solutions*, are not new for our community. Many folks in our diaspora lack consistent access to affordable, nutritious, and culturally appropriate foods, but there are very real barriers to government assistance programs like SNAP (Supplemental Nutrition Assistance Program) and WIC (Women, Infants, and Children supplemental nutrition program). A lot of Asian American communities don't go on SNAP or WIC, sometimes because they don't want to or know how, but oftentimes because of shame factors and cultural resilience issues. I grew up with a single mother, and in college I would use my meal points to buy groceries for my family, but we never went on food stamps. For my community, it was not something that was ever in the consciousness. The prevailing sentiment was, I don't want to be counted by the government, because the last time folks were counted by the government, they were sent off to war, or they were hauled off to internment camps. Instead, we navigated food access by saying, so-and-so's aunt grows this, and if we cook together once a week, we'll be able to have x, y, and z meals together. The less social connection that we had, the hungrier I felt as a child, and vice versa.

On the need for radical intersectionality in the environmental justice movement

For me, environmental justice is reproductive justice is racial justice. As a social worker, I've seen how intimate partner violence spikes after hurricanes. We see people get stuck in a place where they breathe in an inordinate amount of toxic chemicals and then wonder why their parents are sick or why they're infertile. These issues are palpable and real in our community, and making these connections is essential. People often say that **critical race theory**, which has become so controversial recently, is a component of environmental justice or reproductive justice work. I'm like, no, that's the *foundation* of this work. That has been the fight.

At the same time, people of color are the prototype in developing innovations and designing justice in these organic, community organizing spaces. It's really sacred but also really unacknowledged labor. For me, so much of decolonizing this work has been about recognizing that Asian Americans have always been in environmental justice; our diaspora has always had social workers and healers and folks who work with the land and the resources that we're given, but we've just called them different names.

On the importance of centering elders

VAYLA does a lot of leadership development, and people often envision that we're working with these pristine fourteen-year-olds when we talk about the leaders of tomorrow. I'm like, no, we're talking about your grandma! It's very presumptuous to assume that they won't be around for a long time! I like to call ourselves intergenerational mediators. Our elders are often the most impacted by injustices. A lot of the older folks we work with are also spectacular storytellers. They have lived through so much and uphold so much medicine in their existence.

JANELLE ST. JOHN

SHE/HER/HERS

"Urban agriculture shouldn't be a nice-to-have—it's a need-to-have"

Janelle St. John knows that local food production and food sovereignty are cornerstones of community resilience, particularly amid an increasingly volatile climate, and that's why she's stewarding an urban farming movement in Englewood, a predominantly Black neighborhood in Chicago's South Side. As the executive director of Growing Home, a USDA-certified organic, high-production farm and job training social enterprise, Janelle is eager to continue expanding affordable, healthy, and desirable food access and well-paying job opportunities for her neighbors and to grow the farm into a campus that can serve as a community anchor for folks to gather, connect, and spend time with their hands in the soil.

Janelle brings to the Growing Home team decades of experience honing her nonprofit development and management expertise across community centers, Boys and Girls Clubs, children's foundations, mental health organizations, and more. As exemplified by her storytelling chops and ability to rattle off innumerable statistics from memory, Janelle is highly skilled in cultivating partnership and deep investment in communities that have historically experienced disinvestment. While she is relatively new to the realm of urban agriculture, she innately understands the critical importance of local food sovereignty after growing up in rural Trinidad and Tobago, where she watched once-robust local food production decline and food costs rise in tandem with the oil industry boom. She is a lifelong and effective servant leader, leading with equal parts heart, lived expertise, and data.

Content Notes: Food apartheid, allusion to mass incarceration

On urban agriculture as a necessity

When I joined the Growing Home team a couple of years ago, we were selling 80–90 percent of the food that we grew at farmers markets on the North Side of Chicago. Englewood is a **food desert**, and the vast majority of the food we were growing was leaving our community. It was wild! So at the start of 2020, we committed to distributing at least half of our produce locally. The pandemic hit a few months later, and we ended up donating nearly our entire harvests to neighbors. By the end of the year, we had delivered more than 1,200 boxes of fresh food to Englewood residents at no cost.

Covid-19 proved that urban agriculture shouldn't be a nice-to-have but a need-to-have. It taught us that any community can become a food desert. And I don't always like the term *food desert*, because we have grocery stores, but it's about having consistent access to affordable, healthy options. When your food has to travel from 50, 100, 200 miles away, anything that interrupts that process can prevent you from being able to eat. So should cities be investing in the local agriculture space? Absolutely! When the pandemic hit, we were already here, we had the infrastructure, and we had the expertise. We were already growing 30,000 pounds of food on our urban farms. We were able to immediately pivot to do door-to-door deliveries and make sure that our neighbors had essential food access.

We used to do an average of thirty **community-supported agriculture** (CSA) boxes for a sixteen-week growing season. Now we deliver ninety CSA boxes every other week, and we intend to do so year-round. You see a lot of food giveaways, but they're one week here and one week there. For me, it's important that the same families can look forward to getting high-quality food on a consistent basis.

So many people have a perception of a need and they come into a community and build something that nobody uses. They think, *This Black community should want a bike path, because I have a bike path*. But you know what? I don't want a bike path if I'm going to be robbed. Safety is bigger than a bike path. Similarly, I accept that produce isn't the top thing on most people's lists. But I also understand that access to fruits and vegetables is going to affect the health outcomes in our community. So my approach is, *How can I make sure that everyone has access to fresh, healthy food, understanding that they have other priorities?*

On the ripple effects of Growing Home's job training program

Part of what we do at Growing Home is create pathways to sustainable careers and provide incredible job training. Since 2002 we've been a leader in farm-based job training and job training for folks with barriers to employment. The results have been remarkable. (In a study of Growing Home's alumni between 2016 and 2018, researchers found that the Growing Home job training program reduced the likelihood of recidivism by 71 percent compared to the Illinois state average!)

Each year, we've supported an average of about fifty-five program assistants, which is what we call our job trainees, but we're currently working on expanding the program to include an additional cohort. When I say that we train fifty-five individuals, we're really training fifty-five families. Most people don't get that. When we train people to grow food, they're taking that home. They're growing fresh food in their yards, they're sharing their new skills and their excitement for fresh produce with their loved ones, and they're coming home calmer and more confident. That's how you impact the community.

Many of the program assistants find the farm experience to be therapeutic. In fact, 78 percent of the program graduates reported an improvement in their health and mental health after completing the program. Many of them have been in the jail or prison system for years. Many of them are veterans. Being outside, on the land, with their hands in the soil, is healing.

On community as motivation

I grew up on the North Side of Chicago after moving here from Trinidad and Tobago. I would never go past 95th Street. As a North Side Black girl, the South Side was a blanket place of fear and violence. I knew nothing about the South Side. When my sister moved to Englewood, I saw this amazing community where everyone owned their own home and mowed their lawns. People never talk about that! Now that I live here, I'm the youngest person on the street because it's mostly older folks, and I feel like I'm on the set of *Leave It to Beaver*. My kids will go out on the street to play, and I'll hear a neighbor yell, "Don't worry, Ms. Janelle, I'm watching them." My kids and my family have been so embraced by the community.

That's what motivates me. My community is amazing. They deserve to have good schools and good homes and good food. When people say Englewood is this, Englewood is that, I want to hear people say, "Oh, but did you hear that they have their own USDA-certified organic farm? Yes, farming, in Chicago, on the South Side." I want to change the narrative of our community a little bit. That's why I see the impact of our work so far beyond the day-to-day of what we do.

RESILIENCE TOOL SPOTLIGHT

Urban Farming

Urban agriculture has existed as long as cities have existed, though there are a handful of moments in urban agriculture history that are particularly well known; during World War II, in a government-driven push for solidarity with soldiers on the warfront, close to twenty million families planted seven million acres of victory gardens across the United States, in their yards, city hall lawns, and public parks, producing about 40 percent of the vegetables consumed nationally.[1] In the aftermath of the Soviet Union's collapse, as Cuba experienced a major economic crisis and was cut off from trade with the rest of the world, Cubans looked to ancestral farming traditions to transform their food system, growing food in pots on decks, containers on rooftops, and slices of vacant land in town and cities. Today a robust patchwork of 86,000 acres of land comprises the impressive urban agriculture landscape of Havana.[2] In post-industrial Detroit, Michigan, the city's so-called Renaissance is characterized in part by the proliferation of gardens and farms sprouting up across the city's estimated 100,000 vacant lots.[3] The city isn't new to urban agriculture, and in fact urban farming was popularized in Detroit as early as 1893, when the mayor urged starving residents to get through the economic crisis by growing potatoes in their yards and on the city's public land.

During moments of crisis, urban agriculture has been a lifeline. As extreme weather intensifies and becomes more frequent, therefore impacting the production, processing, and transportation of food, growing food close to home can help ensure that people are fed no matter what. Currently just a handful of corporations control the vast majority of the food supply chain in the United States. More than half of fruit and about a third of vegetables consumed in the United States are imported.[4] Since the food system is so concentrated and food supply chains are often thousands of miles long, a small breakdown, whether a power outage, crop loss, or highway closure, can rapidly wreak havoc on the entire system. The more that food systems can decentralize and the closer that food can be grown to home (or more specifically, in cities, where an estimated 82 percent of the American population resides[5]), the better prepared our food systems will be to endure unpredictability and climate chaos.

There are plentiful other benefits to growing food in cities, too. Gardens and farms help absorb rainwater, filter local air pollution, supply habitat for birds and insects, and provide temperature relief within urban heat islands. Rooftop gardens help insulate buildings and lower energy demand. Urban agriculture spaces can be sources of community building, modest income, beauty, mental and physical health benefits, education, and training. Homegrown food picked at peak ripeness is also generally tastier, more nutrient dense, and far more affordable than produce shipped from thousands of miles away. From community orchards to vertical farms, plot-based community gardens to backyard chicken coops, rooftop beehives to aquaponics greenhouses, the urban agriculture tapestry is diverse and bountiful. There is no singular urban farming playbook, so there's tremendous freedom to figure out what makes the most sense for every community.

Urban agriculture is not a panacea, though. City gardens and farms have repeatedly been linked with gentrification and displacement, particularly when the efforts are led by community outsiders or co-opted by developers. The urban farming ventures that receive the most funding often do the least to challenge the root of exploitative food systems and food injustices. Growing enough food to meaningfully feed city residents requires vast amounts of space, which cities rarely have in abundance, and many argue that land would be better utilized for housing rather than instigating further sprawl. Lots of urban farmers themselves lament that the benefits of urban agriculture are overhyped and oversold. Zoning and public nuisance laws often criminalize or complicate the growing of food. And while growing food can be an act of self-determination and liberation, many communities, particularly the African diaspora, have experienced a history of exploitation and enslavement for agricultural labor. Growing food in cities is powerful and it is also complex. Like all resilience tools featured in this book, it's just one tool, and it should be approached with nuance and care.

To get involved with the urban agriculture movement in your community:

- **Research prompts:** If you're a city dweller, what urban farms already exist or are emerging in your area? Which efforts are community-driven and committed to advancing liberation and justice? Which organizers are working hard to advance the urban agriculture movement in your community? If nearby grocery stores couldn't stock produce for weeks or months, where might you be able to find basic necessities elsewhere in your community? If you can't find any urban agriculture projects in your area, might zoning restrictions, public nuisance laws, or other restrictive policies be a factor?
- **Journal prompt:** Imagine if city streets, parking spaces, and vacant lots were transformed into gardens and farms. What would that look like, feel like, smell like, taste like?
- **If you enjoy regularly cooking:** Consider joining a values-aligned, community-supported agriculture (CSA) food box program, which is how many urban farms financially sustain their options. CSA boxes

are typically delivered on a weekly basis and include a mix of seasonal vegetables and fruits, sometimes with the option to add other delicacies made on the farm. A CSA box subscription is a fantastic way to consistently support urban farmers and ensure that your kitchen is always stocked with delicious produce. Some CSA programs have sliding scale pricing or offer subsidized or no-cost boxes for low-income folks.

- **If you have access to time and sunlight:** Begin growing food at home. Nutrient-dense sprouts and microgreens can be grown on kitchen counters with a sunny window. If you have access to a deck, courtyard, or rooftop that receives several hours of sunshine each day, try growing greens, root vegetables like carrots and radishes, citrus, and herbs in pots, growbags, or reused containers. If you have access to a yard, try growing a kitchen garden in raised beds (or plant food in-ground, but make sure to test your soil first). According to the *Old Farmer's Almanac*, just 200 square feet is needed to grow enough produce to feed an omnivore for a year! That being said, start small, go slow, and share your bounty with friends and neighbors when the harvest is overwhelming.
- **If you have access to time and a nearby community garden:** Do some research on the community garden that's the best fit for you, and get involved. Would you rather have your own plot to tend on a regular basis, or would you prefer to occasionally join community workdays at a collectively managed garden? If there aren't any community gardens near you, and you have the passion and bandwidth to take on a major project, consider starting one yourself (ideally with other nearby pals).
- **If there are policy-related barriers to urban agriculture in your community:** Organize with friends and neighbors to advocate for policy that supports and decriminalizes growing food in cities. Get to know the Right to Farm laws in your state. Look to cities with robust urban agriculture landscapes for policy inspiration. Contact your

local representatives and attend city meetings to express why it's an important issue to you and what specifically you'd like to see change.

- **If you work in city planning, architecture, development, or design:** Create spaces and policies with the intention to enable the growing of food on private and publicly owned lands. Partner with organizers and listen to community members to learn more about the barriers that currently exist and better understand the solutions that they'd like to see.
- **General best practices:** Befriend your local nursery and lean on their team for guidance. Test for soil contaminants like lead, arsenic, and other heavy metals before planting food in-ground. If you have the opportunity, get to know sunlight exposure at different times of the day and the year before selecting a place to plant. Start small and harvest regularly. Remember to add plants that will invite pollinators to your garden. In any efforts beyond your home, follow the lead of and center the community members who are most impacted by food injustices and would be most vulnerable to potential gentrification. Grow food with others. As guerilla gardening advocate Ron Finley says, "When we grow together, we grow together."

MIRIAM BELBLIDIA

SHE/HER/HERS

"The people who are most impacted know best what they need"

Miriam Belblidia is a leading water-management and hazard-mitigation strategist and organizer based in Bulbancha, presently known as New Orleans, Louisiana. Miriam worked as the director of programs at Imagine Water Works, an organization that she cofounded in 2012 as a social enterprise dedicated to multidisciplinary flood mitigation and water quality solutions. Over the years, Imagine Water Works has evolved to center QTBIPOC folks, lift up local expertise, and merge science with art, history, connection, and joy. In the last few years, as many Louisiana communities have endured the impacts of brutal hurricane seasons and the pandemic, the organization has led powerful relief, recovery, and preparedness efforts rooted in reciprocal and collective care. In fact, I first got to know Miriam when we collaborated on a digital guide to preparing for hurricane season. Just a few days after we shared the project, Hurricane Ida hit, and I watched in awe as Miriam and her colleagues led a massive mutual aid effort, disbursing tens of thousands of dollars and distributing supplies on the ground. Their team is nothing short of incredible.

Beyond Imagine Water Works, Miriam cofounded the Water Collaborative of Greater New Orleans, a regional water management group of more than 300 members. She served as the interim director and visiting senior lecturer of her alma mater, the Center for Disaster Management at the University of Pittsburgh. She has also published influential pieces of scholarly literature like "Building Community Resilience through Social Networking Sites" and "Mutual Aid: A

Grassroots Model for Justice and Equity in Emergency Management." Miriam, who is a queer Algerian American woman, brings an intersectional, grassroots lens and infectious zeal to all that she does.

Content Notes: Mention of structural and interpersonal anti-trans discrimination

On the importance of centering QTBIPOC in climate justice work

I've been in New Orleans more or less since 2009, which is the longest amount of time that I've lived anywhere. Still, I'll never be from Louisiana. I came here specifically for work as part of the post-Katrina influx. My presence is complicated. Part of what I value about New Orleans is that, as a port town, it's a place that has always been in flux, from the natural environment to the flow of people. Bulbancha, the Choctaw name for the area, literally means "place of many tongues." As a person who spent her childhood moving from Atlanta to Algeria to Switzerland to Maryland, I cherish living in a place that has long been very fluid. It's important to recognize, though, that this fluidity comes with displacement and gentrification and other harmful cycles.

Around 2017 I began thinking deeply about whether or not I was the right person to be leading this work. I am a white-presenting transplant and there are plenty of us in the nonprofit world, especially in New Orleans and the world of water. The whiteness of this community is especially egregious given the demographics of the city.

So I stepped down and Klie Kliebert, a native New Orleanian and trans person who had joined our team as a communications intern a couple of years prior, stepped into the role of executive director. We expanded our team and shifted to *Imagine* Water Works (from Water Works previously), because we wanted to focus on the creativity and imagination required for building what we wanted to build.

Klie being a native New Orleanian is crucial to the success of this work. Deja Jones, our programs associate and resident artist, being a native New Orleanian is crucial to the success of this work. Part of that is just in their ways of being and building community. There are a lot of mutual aid projects that have popped up, especially during Covid, that were run by transplants, and those projects weren't sustainable because people burnt out or didn't have community ties. It's just really hard to sustain the work if you're not committed to a place.

The shift in leadership has also illuminated how important it is to have trans folks leading not just in trans organizations but in all organizations. For instance, our Trans Clippers Project came from a request for a haircut in one of our mutual aid groups. A lot of people were shaming them because it was the early days of the pandemic and people said that a haircut wasn't a priority. It turned out that this was a trans person, and to maintain their gender identity they really fucking needed a haircut. That's when having trans leadership really matters. Klie could tell that there was more going on than was immediately visible to people. And so we launched this program to deliver electric hair clippers and trimmers to trans and nonbinary folks who need access to hair care to remain safe and healthy during and after disasters. The Trans Clippers Project has since been replicated in thirty-something states, Canada, and Australia.

One of the main grounding principles of our work is trusting people to know what their needs are and not allowing any shame. Only you know what your needs are, and we can't police that! I think that's what really differentiates mutual aid work from emergency management approaches, which have been very militaristic, top-down, "we will prescribe for you what you're going to get and you're going to like it." We have taken an approach that the people who are most impacted know best what they need, and we listen to them. And that has launched a lot of beautiful projects.

On mutual aid in practice

We started our mutual aid work in partnership with Project South and the Southern Movement Assembly.[1] They were working to seed this network of liberation hubs throughout the Deep South where folks could practice mutual aid.[2] We were selected to take the lead in creating mutual aid hubs in Louisiana. So we began cultivating a mutual aid response network which would exist online and in person. We launched our regional mutual aid groups on Facebook in March 2020, the start of the Covid pandemic, and then Hurricane Laura hit, and then Zeta, and then Ida. It's been nonstop.

When Hurricane Ida hit, Imagine Water Works went viral. We were on Barack Obama's list of places to donate! To go viral like that was a blessing and a curse. We brought in half a million dollars from that fundraiser, which is more money than is conceivable to me. And we were moving it pretty damn quickly. Four days after the storm we had redistributed $40,000, and we still didn't even have power. We did a lot of direct distribution to other BIPOC-led groups that we trusted, like the Birthmark Doula Collective and Familias Unidas, who we knew would get the money out to their communities. We were also buying and distributing supplies. But the donations were coming in at such a clip that there were limits to how quickly we could move money. Platforms like PayPal and Venmo have limits on how much money you can transfer at once. And it takes time to figure out how you're going to disburse funds! You can't just send somebody $10,000 without talking to them. But people saw us bringing in money and wanted us to spend all that money immediately. We started getting backlash online, mostly from folks who were not local or on the ground. I distinctly recall being at a warehouse without power, coordinating supply distribution while trying to navigate waves of backlash on the internet. It was a nightmare.

I'm proud of us for handling the situation and not caving to the public pressure, though. We do this work based on our experience, and we try to be really intentional and not reactive in our approach. That's difficult because people are incredibly reactive and agitated during a disaster. But in that moment, we knew that we had more money than we'd probably ever see again, and so we decided to divide it into thirds: one-third for immediate response, one-third for recovery needs, and another third to prepare for the next thing. There is always an influx of resources—cash, food, toiletries, and volunteers—in the aftermath of a disaster when everyone wants to help, but that dies down really quickly. The groups that aren't based locally pull out. There are often no resources left for recovery and preparedness. We didn't want to fall into the same trap.

Being part of mutual aid efforts on the ground and in person was such a beautiful experience compared to online, and that was interesting to witness. Mutual aid has taken off as a concept in the last few years in ways that I think are really exciting. But I think some people misunderstand the roots and the history of mutual aid. New Orleans in particular has a deep history of social aid and pleasure clubs that date back hundreds of years and are rooted in Black and Indigenous ways of surviving and organizing.[3] We try very hard to root our work in those traditions. Mutual aid is not just cash disbursements and screenshots of Venmo donations. It can be a part of it. It's important. But it's not the end-all, be-all. In the end, mutual aid is about strengthening community connections.

RESILIENCE TOOL SPOTLIGHT
Mutual Aid

Mutual aid is perhaps one of the most powerful tools that we have at our disposal to keep our communities safe, nourished, and wholly cared for through both the slowly simmering harm and major disasters that the climate crisis and interrelated crises bring.

As organizer Dean Spade explains in his 2020 book *Mutual Aid*, "Mutual aid is collective coordination to meet each other's needs, usually from an awareness that the systems we have in place are not going to meet them . . . This survival work, when done in conjunction with social movements demanding transformative change, is called mutual aid."[4] Mutual aid is distinctly different from charity or aid models, which generally involve wealthy individuals or government entities determining what services should look like and who is eligible to receive them. Charity and aid programs typically operate from a deficit-based perspective, rarely confront systemic issues, and are typically exceedingly insufficient if not actively harmful. Mutual aid projects, on the other hand, strive to radically redistribute resources and care while mobilizing community members, building shared understanding, and working to address the root of injustices. Projects are determined by community members and are rooted in a profound recognition of shared humanity and the abundance already present.

While the term *mutual aid* is relatively new, coined by nineteenth-century anarchist Peter Kropotkin after he witnessed animals of the Siberian wilderness collaborating with one another rather than competing for survival, the practice is ancient. Resource sharing, reciprocity, and collective care are ancestral concepts with Indigenous roots. Despite capitalism and colonialism disrupting these millennia-old practices, countless communities have found ways to keep networks of mutual aid alive to meet everyone's needs through acute and long-term crises.

Beginning in the late 1700s, Black communities across Turtle Island pooled resources to care for the sick, support widows and orphans, bury the dead, and provide education, health insurance, and life insurance. Similarly, in the nineteenth and twentieth centuries, Latine communities in the American Southwest formed sociedades mutualistas to create alternatives to the public and private services that they were excluded from and to organize for higher wages. In the 1960s and 1970s, mutual aid projects were the backbone of many movements. Perhaps most notably, the Black Panther Party created dozens of survival programs, like free breakfast, education, medical clinics, rides for elderly folks, and more, where people could come together to get their basic needs met while building a shared analysis, forming meaningful relationships, and strategizing for action. The Young Lords Party for Puerto Rico liberation utilized a similar approach, and even occupied hospitals, churches, and vacant land across Chicago to care for their people and make demands of the city. Beginning in the 1980s, as the state failed to respond to the HIV/AIDS epidemic, queer communities mobilized to create health clinics and provide practical and emotional support to one another while organizing to inform the public, fight discrimination, and demand treatment. Similarly, in the past few years, as the government failed to adequately respond to the Covid-19 pandemic, anti-Black racism, and police violence, mutual aid networks proliferated and gained mainstream recognition. Neighborhoods and affinity groups have banded together to help one another pay rent, get groceries, cover protestors' bail, and more.

As Dean Spade says, "At its best, mutual aid actually produces new ways of living where people get to create systems of care and generosity that address harm and foster well-being . . . Mutual aid projects let us practice meeting our own and each other's needs, based in shared commitments to dignity, care, and justice."[5] The more that we practice, the better positioned we will be to respond to the next hurricane, legislative attack, or pandemic.

To get involved with mutual aid efforts in your community(s):

- **Research prompts:** Make a list of the communities that you're a part of. How have they historically practiced collective care, resource sharing, and reciprocity? Are there any mutual aid efforts underway in communities that you share a common experience or location with? Platforms like www.mutualaidhub.org and www.itsgoingdown.org are great places to begin your search, as are the newsletters from grassroots organizations, the social media accounts of local organizers, and the bulletin boards at local collectives and cooperatives.
- **Journal prompt:** How might you incorporate mutual aid principles into the communities and movements that you're already engaged with? What do you have in abundance that you can share, and what are you lacking or needing that your community might be able to help provide?
- **If you have access to time and a passion for food:** Consider getting involved in a local food sharing initiative, like a community fridge or pantry. There are countless roles needed to keep community fridges operating smoothly, so inquire about the type of support needed and consider what makes the most sense for you. This might look like signing up for a weekly slot to clean out the fridge, dropping off groceries or produce from your garden on a regular basis, offering to liaise between local farms or bakeries to transport high-quality donations, or cooking large batches of your favorite nourishing dishes to stock the fridge with premade, individually portioned meals. If you have the bandwidth, you might even consider starting a new food sharing initiative in your community. Online sources like www.freedge.org offer comprehensive guides to building a community fridge program. As always, don't go it alone; involve as many community members as possible, and try to connect with folks who have done it before themselves.

- **If you have access to financial resources:** Move money to mutual aid efforts on a regular basis. This may look like setting up recurring donations online or holding yourself accountable to a weekly practice and transferring funds directly to individuals or groups that request financial assistance via social media.
- **If you have access to a robust local network:** Activate your people to tap into existing mutual aid networks. Gather your pals to attend an orientation or meeting together and hold one another accountable to show up consistently when your support is needed. Or act as a liaison between mutual aid hubs and folks in your community in need of certain resources. Help distribute goods or connect folks to support during and after disasters.
- **If you have shareable skills:** There are a multitude of types of mutual aid initiatives and therefore a multitude of different skill sets and passions needed. Whether you practice acupuncture or love to organize logistics, have a background in law or can translate conversations, design beautiful graphics or are a fantastic listener, there is a place in the constellation of mutual aid efforts where your gifts will be deeply appreciated.
- **General best practices:** Dean Spade explains that "mutual aid groups face four dangerous tendencies: dividing people into those who are deserving and undeserving of help, practicing saviorism, being co-opted, and collaborating with efforts to eliminate public infrastructure and replace it with private enterprise and volunteerism."[6] To resist these pitfalls, make sure that you're building participatory, transparent, and intentional decision-making structures from the get-go, and check out Spade's *Mutual Aid* for a comprehensive guide to doing so. Additionally, keep in mind that mutual aid works best when projects are relatively small (scaling up and out should not be the goal), place-based, and decentralized, but that doesn't mean we can't learn from and support other projects; the Trans Clippers Project being adopted in dozens of other locations is a perfect example of that. Imagine

Water Works was transparent with their efforts so that their processes could be easily replicated and customized elsewhere. Prioritize fun, connection, authenticity, and space for healing to minimize burnout, attract new folks, and ensure that existing folks can stick around for the long haul. Some of the most well-known mutual aid efforts today, like For the Gworls, a Black-trans-led collective that fundraises for their community by curating epic parties, have flourished precisely because they have made joy, art, relationship, and sense of belonging central to their efforts.

LIL MILAGRO HENRIQUEZ

SHE/HER/HERS

"I wanted young people to know that their people have survived for generations in sustainable, regenerative relationship with the land"

Lil Milagro Henriquez is the founder and executive director of Mycelium Youth Network (MYN), an organization dedicated to preparing and empowering young people of color for climate change through place-based ancestral knowledge, storytelling, and holistic, hands-on STEAM (science, technology, engineering, arts, and math) programming. The seeds for MYN were planted in 2017, as flames engulfed California in one of the state's most destructive wildfires to date and Lil Milagro, who had recently given birth, ached with anxiety about the world she would be leaving her children. She sought out disaster-preparedness courses but found that there were few options, and even fewer options that were justice oriented, rooted in community, and accessible to young people. MYN was created to fill that gap. Today the nonprofit offers climate resilience curricula for schools, afterschool programs, and affinity spaces for Native, Black, and queer and trans youth. I've had the privilege of getting to know MYN more deeply in the past few years, and I hope that one day far more young people will have access to similar offerings in their communities.

Beyond her role as MYN's executive director, Lil Milagro is a proud sci-fi fantasy nerd, liberation theologian, and seasoned organizer, having spent decades working on access to higher education for low-income and communities of color, food sovereignty, environmental racism, union democracy, and labor rights. She is also a detribalized descendant of Nahuatl-Pipil people from Cuscatlán, now known as El Salvador. She serves on steering committees for the International Transformational Resilience Coalition, the Oakland Climate Action Coalition, and *Estuary* magazine; is a strategic advisor of Nature-Based Climate Initiatives; and is a recipient of the Women's Earth Alliance fellowship. In every role that she fills, Lil Milagro brings an anti-oppressive, feminist, and union-organizing perspective and is motivated by making her children and ancestors proud.

Content Notes: Environmental destruction and injustice, settler colonization

On her family's history with environmental injustices

Climate change and environmental injustice have been at the root of so much of my family history. When I think about Cuscatlán, currently known as El Salvador, where my mom is from, approximately 98 percent of its old-growth forest has been wiped out. It appears like there are still many trees and tons of vegetation, but that can be deceiving. Those trees were largely planted for shade-grown coffee, as El Salvador was once one of the biggest exporters of coffee in the world. I have a distinct memory of working on my master's thesis in El Salvador and observing the mountainside next to my uncle's home. It looked really green, but when I looked closer, it was all just corn. Every single time that it rains in that part of El Salvador, there are huge mudslides. Because all the old-growth forest is gone, there's nothing holding the soil together. The landscape has been stripped of all the components that would act as a buffer against massive amounts of rain or hurricanes.

In New Orleans, where my parents met and where I grew up, Hurricane Katrina hit shortly after I left for college. I remember trying to convince my father that it was going to be serious enough that he had to leave. Of course, nobody really knew how bad it was going to be. It took me *months* to find out that he was okay—he doesn't really believe in cell phones and cell phones weren't as prevalent at the time. After that, it took him months and months and months to rebuild. The lack of government care for the people that were there was astounding. He was given a FEMA trailer with mold and asbestos. But his neighbors in New Orleans East threw in and supported him. When he was able to rebuild, it was because of the community that came through.

So when I think about issues of environmental concern for my communities, I recognize that so much of my family's history has already been tied to environmental injustice, environmental racism, environmental change, and colonization. High levels of environmental destruction were underway years and years before I was even born. As we rapidly approach a moment in human history when climate change may push us over a climate cliff, we know that environmental injustices will be further heightened by climate change and extreme weather. I am terrified by what that means for my community's ability to survive. I'm also deeply, deeply hopeful about what happens when we are able to pull in our ancestral traditions to create resilience, solutions, and a different way of being in relationship to earth.

On the importance of empowering youth to navigate the climate crisis

I started Mycelium Youth Network after watching the Tubbs Fire unfold. (The Tubbs Fire, which swept Napa and Sonoma counties in October 2017, was, at the time, the most destructive and deadliest fire in the state's history.) I remember thinking about all the ways in which young people weren't being prepared for climate change. There are so many organizations doing amazing environmental justice and activism work, but I saw a need to directly empower young people. I wanted young people to know that their people, their history, and their lineages are incredibly strong and have survived for generations upon generations in sustainable, regenerative relationship with the land. I also wanted to make sure that they had this toolkit of practices to access if they wanted to.

I knew that I wanted to pull from that place of strength and resilience that is omnipresent in our community, rather than starting from a place of trauma. There's already so much discussion around the trauma that young people hold and the trauma that "underresourced," "underserved," "disadvantaged," or minority communities hold. All those words mean *less than*. They're psychologically impactful. When I look at our community, that's not what I see. I see a community that has gone through 400-plus years of colonization and forced assimilation, has had to leave their homelands and consistently rebuild, has found new ways to pass on legacy and tradition, and is creating so much beauty. Therefore, I knew that I wanted our programs to reflect the strength, beauty, and resilience of our communities. It felt really important to start from a place of ancestral traditions and practices.

On top of that, we layer STEAM programming: more traditionally understood Western science, technology, engineering, arts, and math. And at the core of everything is a love of community, history, and ancestry. We weave it all together with hands-on skills that we believe are necessary to address, like building simple rainwater catchment and air filtration systems, cultivating sacred plant medicine, and utilizing ancestral breathing and mindfulness techniques. That way, if something as impactful as Hurricane Katrina were to happen here, young people would feel prepared to bounce back and respond in a way that feels powerful.

The basis of our Climate Resilient Schools Initiative is a curriculum that informs students about how climate change will impact Oakland and how it has already impacted Oakland. But from there, we allow young people to figure out the areas that *they* want to center in their exploration. They get to choose their own youth participatory action projects to develop. That might look like a gardening project, or an exploration of ancestral history as it relates to composting, or a collaboration with local scientists to map sea-level rise.

On the power of play

When the pandemic hit, we asked young people what they wanted to do and learn. They told us that they're already under so much pressure to perform during a pandemic, so they didn't want anything super academic. That's how our Gaming for Justice program emerged. We're creating environmental Dungeons and Dragons adventures that pull from historical and current topics such as deforestation, sea-level rise, and air pollution, all of which take place in the San Francisco Bay Area.

As an avid gamer myself, I think the power of play is so exciting. I love the ways in which gaming and science fiction fantasy allow us to create a safe space to think about really difficult topics. We get the chance to see ourselves as the heroes, to fail, to win, and to fail some more. It's an opportunity to build social-emotional skills necessary for life and especially necessary for climate change. For young people, their imagination isn't yet boxed in by the constraints of the world. Their creativity is so powerful to see.

RESILIENCE TOOL SPOTLIGHT
Community Science

Community science—sometimes described as citizen science, participatory action research, participatory monitoring, or public participation in scientific research—refers to a model of data collection and knowledge sharing that is community-led and community-controlled. It's intended to enhance local knowledge and understanding, increase scientific engagement and accessibility, help identify solutions, and inspire collective action. It is a powerful way for community members to monitor how the climate crisis and intertwined crises are impacting their surroundings and to prepare for what's to come. It's a particularly potent tool in areas where climate change denialism is pervasive, as it enables folks to track changes firsthand and to explore issues most pertinent to their lives and interests. Community science also provides a gentle opportunity to deepen one's connections to the nearby community, including human and more-than-human members.

While community science has surged in popularity in recent years, particularly as the internet has made it easier to crowdsource data, the origins of the community science framework go back at least a couple of millennia; there's evidence of residents of ancient China tracking migratory locust outbreaks to help inform agricultural practices more than 2,000 years ago.[1] This method of knowledge co-production differs in a handful of ways from scientific research conducted exclusively in academic and government spaces:

- Community science may include Western scientific research methods but can also draw from lived experience, ancestral knowledge, and community-identified solutions. Community science recognizes multiple ways of knowing as valid and important.
- Community science projects oftentimes explicitly center liberation, reclamation, and justice and are rooted in collaboration, consent, co-learning, and self-determination.

- Community science invites community members to cultivate sustained relationships with the world around them and to better understand the inner workings of the ecosystems in which they live. The practice of regularly collecting, making sense of, and communicating data also helps hone critical thinking, scientific literacy, community engagement, and activism skills in an accessible way.
- Rather than studying communities as a problem to be solved or dealt with, community science is a human-centered approach to problem solving that works with community members to identify and mitigate specific points of harm. Community science resists exploitation by scientists and the scientific process.
- Community-led and community-controlled solutions are a natural extension of community-led and community-controlled scientific projects. Rather than relying on interventions to be determined and implemented by academic institutions and governmental agencies, who have been and continue to be responsible for tremendous harm to structurally oppressed communities, community science enables community members to identify best practices themselves.
- Far more data points can be collected in community than by singular scientists. The Global Biodiversity Information Facility has accumulated billions of data points collected from community scientists, informing more than 7,000 peer-reviewed papers.

To get involved with community science efforts:

- **Research prompt:** Are there any community science projects in your area that interest you? Global Biodiversity Information Facility, SciStarter, and CitizenScience.gov are among the several databases of community science projects worth exploring.
- **Journal prompts:** Which modes of knowing and understanding the world were most important to your ancestors? Which modes of knowing and understanding the world around you resonate the most with

you? Which phenomena, issues, or beings in your community would you like to better know and understand?

- **If you're passionate about the weather:** Tap into the ISeeChange network. Simply observe weather and climate near you and describe what you're observing and how it's affecting you, and the platform creates a community record syncing with weather data and trends, which will help guide the work of climatologists, social scientists, urban planners, engineers, and health professionals.
- **If you're passionate about the more-than-human members of your community:** Explore social networks like iNaturalist, which allow you to share photos of flora and fauna for community identification and discussion and can help track changes in the surrounding ecosystem. iNaturalist also has a Projects feature that allows people to pool their observations with others. Consider starting your own or join an existing project in your area.
- **If you're passionate about something specific in your community:** Begin monitoring it on a regular basis, on your own or with neighbors or friends. Perhaps you've been noticing a shift in weather patterns or the physical environment that you'd like to follow more closely, or there's an environmental injustice occurring in your community that you'd like to have more robust data to share, or there's a plant or animal relative that you would like to deepen your relationship with. Consider how much time you have to allocate to this project and what monitoring cadence makes sense for you (a few minutes once a month can still yield meaningful data, depending on the project). Keep a notebook or spreadsheet and jot down what you can observe with your senses. Many additional data points, like air quality, temperature, and GPS location, can be collected via smartphone apps, and simple tools like measuring tape can be helpful, too. Any qualitative or quantitative data that you can make note of is valuable, and there's also tremendous value in the practice of monitoring itself.

- **If you're an educator:** Integrate community science projects into your curriculum. Invite students to formulate their own questions and design their own studies, data collection, and analysis processes. Partner with local organizations, community associations, and grassroots organizers to have students share their findings.
- **If you're craving a creative outlet:** Consider using art to process and express the findings of a community science project. Contemplate what medium might be the best fit for you or your group's skills, interests, and resources, and then think about how you might communicate findings with the intended audience. Murals, public art installations or pop-ups, community art gatherings, and social media or web projects can all be impactful media.
- **General best practices:** Community science is best conducted, you guessed it, in community. Start small. Honor all types of expertise. Build respectful relationships with the human and more-than-human community members you encounter. Do no harm.

Community Adaptation

In the years ahead, communities will need to diligently plan for change and make adjustments, some more drastic than others, in order to keep their people safe amid ever-evolving and uncertain conditions. These essays are from researchers, planners, and community leaders who are cocreating blueprints for community adaptation that optimize for justice and center the lived expertise of those most impacted. They demonstrate how to thoughtfully act upon an understanding of existing and expected climate risks to prevent harm and ensure that all community members can survive and thrive in the future.

AMEE RAVAL

SHE/HER/HERS

"We can shift power when we appropriately value lived experience and knowledge as expertise"

Amee Raval is laying the foundation for more holistic climate resilience and adaptation work in so-called California. After reading some of Amee's opinion pieces and attending webinars spotlighting her impactful work, I was struck by her deeply human, nuanced approach to policy and research. With a scientific background in the public health space, experience as a social needs advocate, and lived expertise as the daughter of immigrants, Amee recognizes the value of both quantitative data and storytelling, science-driven and soul-driven strategy, state policy and grassroots organizing. In everything that she does, Amee invites complexity, embraces the contradictions inherent to complex truths, and creates spaciousness for "both/and" thinking.

As the research and policy director of East Bay–based nonprofit Asian Pacific Environmental Network (APEN), Amee is focused on building people power, uplifting the wisdom of **frontline and fenceline communities**, and keeping her East Bay and South Asian diaspora communities safe through the climate crisis and interrelated crises. She was also recently appointed to California's Environmental Justice Advisory Committee, which advises the California Air Resources Board on its Climate Change Scoping Plan updates. Amee spends her time community organizing, conducting and communicating cutting-edge research, helping inform and pass state policies that challenge the extractive economy and transform the energy system, and collaborating with others to develop regenerative economy solutions.

Content Notes: Mention of miscarriage, environmental injustice, structural racism

On the environmental injustices that inspire and fuel her work

My parents worked in the hospitality industry as I was growing up, saving up toward owning their own motel. It's deep, grueling, physical labor. In addition to arduous physical demands, the housekeeping aspect involves consistent exposure to toxic cleaning products and chemicals. It has meant that my parents have experienced a range of chronic health impacts. To share a deeply personal story, my mom had two miscarriages before I was born. She was handling cleaning products late into her pregnancies, scrubbing the hotel room bathtubs with bleach. Looking back, she says, "I didn't know any better." I hold that in my heart. My parents' stories opened my eyes to the fact that your daily workplace can have a significant impact on health and well-being.

My parents immigrated from India thirty years ago, which inspired my early work as a climate advocate confronting the occupational health impacts of extreme heat in our hometown. Parts of South Asia are expected to become uninhabitable as global temperatures rise and those localized heat impacts become unbearable.

My personal experiences, growing up as a child of immigrants and connecting to my homeland, have culminated in the work that I currently do with the Asian Pacific Environmental Network, advocating for Asian American immigrant and refugee communities here in the East Bay. My parents sacrificed so much to give my brother and me opportunities and I feel like I'm carrying on their legacy through my advocacy work. I carry with me all that they fought for to come here and fought against once they were here. I work for the hope and the vision that families like mine don't have to sacrifice for their health and their well-being.

In Richmond and Oakland's Chinatown, there are historic, place-based environmental and health burdens that have already taken their toll. In Richmond our community members are living on the fenceline of the Chevron oil refinery. In Oakland's Chinatown our members describe the neighborhood as a wok of pollution. Surrounded by three freeways and a shipping port, pollution ends up sitting, stagnant and heavy, in their neighborhoods. Given the magnitude of trucks traversing those freeways, they're impacted by diesel pollution in particular. As a result, these communities experience higher asthma rates.

These existing impacts are now being compounded by climate disasters. Our refugee and immigrant communities are facing both the historic impacts of climate pollution as well as worsening climate impacts across the Bay Area, such as power shutoffs, fires, toxic smoke, and heat waves.[1]

On mapping community climate resilience

One of the areas where APEN is currently focusing is this concept of resilience mapping. To better understand which communities are most impacted by climate change, we know that climate science alone isn't enough. It doesn't tell us the full story of who's disproportionately impacted. For instance, when considering who will be most impacted by extreme heat days, it's not enough to map local temperatures; we must also consider how heat overlays with socioeconomic and neighborhood conditions. Those are the elements that determine whether you can afford air-conditioning, if you have easy access to cooling, whether you're surrounded by green space or asphalt, if you have vehicle or transit access, and on and on. Those are the factors that make or break whether a community can cope and respond. It's one thing for it to be hot outside, but it's another when you live in an older building and don't have ceiling fans or air-conditioning and it's not well weatherized. It's another thing when you open the windows to cool off and it's really polluted outside because the Chevron refinery is flaring.

More holistic methods of resilience mapping are critical because they deepen our understanding of the cumulative impacts of climate change and socioeconomic vulnerability. It's an approach that's largely understudied, but it's really important. It's one way that we can start to depict, legitimize, and visibilize those cumulative impacts.

In efforts to map community climate resilience, it's so easy to be deficiency oriented. Our communities are often characterized as victims, as disadvantaged, as suffering. Yes, there have been historic inequalities that have harmed our communities. And it's important to name that we're also protagonists in our stories. This story is not defined by the Chevron refinery, its pollution, and its chronic health impacts. This story is defined by the power building, the organizing, and the revisioning for a clean, healthy, and thriving community. There's a tremendous richness of social connections and community-based organizations, and it's critical that we uplift those stories, too. When our mapping focuses on vulnerability and doomsday narratives of climate change, these messages can incite a lot of fear. We want to make sure that our mapping captures narratives of hope and resilience as well.

On the need for more resilience hubs and proper valuation of lived expertise

The Asian Pacific Environmental Network is also currently advocating for more resilience hubs. Resilience hubs are anchor institutions that we're familiar with and provide critical services to neighbors every day: schools, libraries, youth centers, parks, recreation sites, and places of worship. In the past few months, as we've lived through intersecting crises, we've seen schools provide food to families and we've watched libraries become cooling centers for unhoused folks during heat waves or to offer clean-air respite during the fires. So while this term is new, resilience hubs aren't a new concept. We've been investing in community and public institutions. However, at this confluence of intersecting crises, these spaces are more important than ever.

Successful resilience hubs require an empowered care workforce. When we talk about our care economy, we're talking about child rearing, we're talking about our social workers, our educators, our counselors, home care workers, healthcare workers. Our care economy is oftentimes invisible, underpaid, and undervalued. And it's responsible for keeping everything going. Our care workers are really meeting the economic and social needs of our communities. Amid the coronavirus pandemic, it's become incredibly apparent just how critical and undervalued our care workers are to keep society functioning. We're advocating for a care workforce that is well trained, well paid, and empowered to act during crises.

To respond effectively to this moment, as we simultaneously confront economic fallout from a pandemic, an ongoing racial justice crisis, and a climate crisis, we need to double down on our investment in public institutions and community-serving facilities, and we need to uplift the deep expertise and storytelling of our community members who have been most impacted by these crises. Traditionally most of the decisions that impact our communities are made by affluent white folks, and particularly by affluent white men. They hold a lot of power in our society and in the mainstream environmental movement. We're arguing that it's working-class communities of color, who are living closest to the problems we're solving for, who are the real experts on issues like economic inequality, public health disparities, and climate change. We can shift power when we appropriately value lived experience and knowledge as expertise.

RESILIENCE TOOL SPOTLIGHT

Community Resilience Mapping

Community resilience maps are a fantastic way to identify and deepen understanding of current and anticipated climate risks and strengths across the community. Climate resilience maps may consider (1) environmental exposures, or the magnitude, frequency, and duration of present and anticipated environmental risks, including both major shocks and slow, simmering stressors; (2) sensitivities, or the physiological, socioeconomic, infrastructural, occupational, and geographic factors that affect the degree to which a population is impacted by climate-related changes; (3) assets, or the factors that affect the degree to which a population is able to bounce back from climate-related changes and care for one another amid crisis; and (4) adaptive capacity, or the infrastructure that may help (or hinder) the mitigation of possible disasters and/or coping with their consequences.[2] Indicators to consider might include the following:

EXPOSURES	ADAPTIVE CAPACITY
Temperature	Vehicle and transit access
Wildfire threat	Medical facilities
Sea-level rise	Mental health care
Flood risk	Cooling/evacuation centers
Drought	Emergency services
Air quality	
Hurricanes	

SENSITIVITIES	ASSETS
Demographic: race/ethnicity, linguistic isolation, poverty, elderly (and living alone), children, disability, incarceration, citizenship status	**Demographic:** folks who speak multiple languages (including ASL), age diversity, experiential diversity
Socioeconomic: unemployment, educational attainment, income inequality, health insurance coverage, food insecurity, voter participation	**Socioeconomic:** community cohesion and engagement, cooperatives, local businesses
Housing and infrastructure: tenure (% renters), affordability (% housing burdened), air-conditioning, quality of housing, energy burden, jails and prisons	**Housing and infrastructure:** community spaces (with industrial kitchens), buildings with energy storage systems, community-based renewable energy systems, local food growing and distribution systems, circular waste management systems
Health: asthma, cardiovascular disease, diabetes	**Occupational:** medical professionals, doulas, herbalists, elder

SENSITIVITIES CONTINUED	**ASSETS** CONTINUED
Occupational: outdoor workers, workers impacted by violations of rights, protections, and safety **Geographic**: urban heat islands, industrial/hazardous sites, impervious surfaces, nearby bodies of water, monocultures	care providers, cooks, domestic workers, educators, facilitators, organizers, mental healthcare providers, informal care workers, carriers of ancestral wisdom **Geographic:** tree canopy, porous/vegetated surfaces, gathering spaces, areas surrounded by defensible space, freshwater sources, gardens, plant medicine, biodiversity

Mapping a variety of indicators is an invaluable way to capture how climate risk and vulnerability may interact to create disasters.[3] For instance, a hurricane is just a natural hazard, not a disaster, until it threatens humans in its wake.[4] However, if we're assessing the risk of a hurricane to a neighborhood that has few porous surfaces, is flanked by rivers and the ocean, and is elevated just above sea level, we can expect that storm surge, torrential rains, and swollen rivers will flood the roads and damage homes. If we also understand that many folks in the neighborhood don't have access to a vehicle to evacuate before the hurricane, lack the financial cushion to preemptively stock up on supplies, and have historically been excluded from accessing emergency services during flooding events, then we begin to understand how hazards become life-threatening disasters. Done well, a resilience map can illustrate an entire cascade of compounding risks.

Thoughtful, comprehensive resilience maps lead to holistic solutions. When resilience and adaptation strategy is informed solely by an awareness of environmental exposures, such as an increased risk of hurricanes and subsequent flooding, solutions tend to focus narrowly on infrastructure, like stronger dams or higher seawalls. But dams sometimes fail, and stormwaters frequently surge past seawalls. Good climate resilience strategies will keep community members safe even when things go awry. Resilience maps that effectively aggregate environmental, physiological, socioeconomic, and experiential data de-silo adaptation efforts and help mitigate compounding risk factors, particularly for those who are most impacted.

To get involved with resilience mapping efforts in your community:

- **Research prompt:** Are there any resilience mapping efforts currently underway in your city, county, or state? What does the public participation process look like? How might you get involved to help represent your lived experience and community?
- **Journal prompt:** Make a list of the environmental exposures, sensitivities, assets, and adaptive capacity of your own household or immediate community. Consider how you might begin to protect and bolster household or community assets, expand your adaptive capacity, ameliorate areas of sensitivity, and prepare for specific environmental exposures.
- **If you have access to a robust community network:** Initiate the process of cocreating a community resilience map with others. Depending on your bandwidth and the scope and intent of your mapping process, this might look like a handful of neighbors gathering in your kitchen around a potluck and a sheet of butcher paper and crayons, or it might be more formal, involving local representatives and a broad swath of constituents coming together at a community center. For more information about how to develop a community resilience mapping process and use it to help inform adaptation and emergency response plans, utilize APEN's report "Mapping Resilience," available as a PDF at www.apen4ej.org,

and explore the *Climate Resilience* companion website www.climate resilienceproject.org for a how-to guide.

- **If you have a few spare minutes:** Contact your local and state representatives to urge them to integrate comprehensive, equity-focused resilience mapping into climate adaptation and community resilience policies and programs.
- **General best practices:** Remember that even the most thoughtful of resilience maps will have gaps. Climate disasters are dynamic and messy and are often exacerbated by risks that haven't been predicted or fully understood. Edit as needed; maps are fluid and will change over time. Resilience maps are important, and they're also just one document.

MARCIE ROTH

SHE/HER/HERS

"You cannot be a resilient community without being an inclusive community, plain and simple"

Marcie Roth has been a lifelong champion for disability rights. I was thrilled to speak with Marcie—and I feel indebted to her—because she has spent the past few decades working especially hard to connect the climate justice and disability rights movements and to meaningfully improve emergency preparedness and disaster outcomes for those of us who are disabled or have disabled beloveds. In 2009 the Obama administration appointed Marcie as a senior advisor for disability issues at FEMA, where she established the Office of Disability Integration and Coordination and led the work of the Department of Homeland Security Interagency Coordinating Council on Emergency Preparedness and Individuals with Disabilities. During her seven-year tenure, she also served as a consultant to the United Nations for the development of disability-inclusive disaster risk reduction principles and served as a leader throughout the development and implementation of the global Sendai Framework for Disaster Risk Reduction 2015–2030.

Marcie currently leads the World Institute on Disability, one of the first global disability rights organizations founded and continually led by people with disabilities, where she works to operationalize disability inclusion and bring policy into action, particularly around accessibility and universal design, community living, and disability-inclusive emergency preparedness and climate resilience. One of her key initiatives has been forming the Global Alliance for Disaster

Resource Acceleration, a collaboration to better connect disaster relief funding streams to disability organizations around the world.

Content Notes: 9/11, ableism, Covid-19 deaths

On disability inclusivity in disaster planning

On September 12, 2001, the day after the terrorist attack in New York City, I was working as the director of advocacy and public policy for the National Council on Independent Living when I got a phone call from a colleague based in New York. He said that there were thousands of people with disabilities in the area around Ground Zero who were not getting the support and services that they needed. That was a lightbulb moment for me: What *does* generally happen to folks with disabilities in disasters? I became really involved in that movement and ultimately was asked to advise the White House on their strategy to ensure that people weren't being denied access to the support and services that they needed. The Obama administration eventually presented me with the charge to establish the Office of Disability Integration and Coordination for FEMA.

During that time, I did a lot of work expanding disaster capacity and transformationally shifting the thinking around disability rights in disasters. However, I didn't immediately connect the dots between my work and climate change. I thought that I was in the civil rights and human rights protections sphere, and climate change didn't fit in that space. It took a while to dial back the work that I was doing in disaster risk reduction and in prevention, preparedness, and planning. Once I made the connection, I wondered why nobody else seemed to notice the siloes.

In reality, people with disabilities are far more impacted in disasters, extreme weather, pandemics, etc. than the rest of the community.[1] It's not because we are more vulnerable, which is a term we are often handed. It's because of the barriers in community life that have yet to be eliminated.

Globally, there are about 1.3 billion people with disabilities.[2] Here in the United States, 26 percent of adults are disabled.[3] We are not an insignificant portion of the population! However, when you think about the impact of climate change on disabled people, not only in moments of extreme weather but also in migration, movement, and disaster planning, the barriers are weighted so much more in the direction of disproportionate harm and exclusion. Among people with disabilities, the folks most impacted are Black people, Indigenous people, people of color, people who experience poverty, LGBTQ+ folks, and other people who are multiply marginalized. They're regularly discriminated against, excluded, and erased.

And yet as a person with a disability myself, and as a parent of two adults with disabilities, I've had to be very good at navigating challenges, emergencies, and disasters. Problem solving is something I get lots of practice at. Not only should I *not* be considered a liability in my community, but I'm actually a tremendous asset when it comes to problem solving and thinking outside the box. I'm really good at having a Plan B, C, or D. So invite me to the table! As people with disabilities, we're not looking for people to plan for us. Plan *with* us. Nothing about us without us.[4]

On the disconnect between disability justice and disaster response

To this day, I think there is a huge disconnect between climate justice work, environmental justice work, and the work that we do in disability justice and rights work. So far, accessibility hasn't been front and center in discussions about disaster and community resilience. There are countless communities that have invested in resilience but haven't given any thought to accessibility. Therefore, they've excluded 26 percent of their community. Resilience for some isn't going to be enough. We don't have the luxury of leaving anybody out of climate disaster and resilience planning. You cannot be a resilient community without being an inclusive community, plain and simple.

This concept, that there is no community resilience without accessibility, inclusivity, and disability justice, has been painfully obvious during the Covid-19 pandemic. Of the people who have died from Covid-19, the majority are referred to as old, vulnerable, or having underlying conditions. Most of those people have disabilities. In nursing homes 94.6 percent of people have disabilities. You don't go to a nursing home because you're old; you go to a nursing home because your community wasn't created to be resilient for everybody. The pandemic has been a horrific erasure of hundreds of thousands of people with disabilities, and primarily Black, Brown, Indigenous, and poor people. We have so clearly failed.

I am not proud of some of the outcomes, I have to say, but I'm cut from the *never give up, never give in* cloth. I'm inspired by the folks who show up each and every day, who continually speak truth to power. There are a few people who are no longer with us who are always in my ear, on my shoulder, kicking my ass. Because the World Institute on Disability has an incredible legacy, I feel both a deep sense of obligation and like I'm surrounded by stardust each day.

I'm also incredibly driven by my granddaughter to keep moving forward. If I'm not getting this work right for my granddaughter, I may as well fold my tent.

EILEEN V. QUIGLEY

SHE/HER/HERS

"There are enormous hazards involved with every aspect of fossil fuels—less dependence on fossil fuels means more reliability over time"

Eileen V. Quigley is a champion for **deep decarbonization** and large-scale transition to clean energy across the American Northwest. She is the founder and executive director of the Clean Energy Transition Institute (CETI), an independent, nonpartisan nonprofit working to advance technical, economic, and equitable decarbonization solutions in Idaho, Montana, Oregon, and Washington. By providing cutting-edge research, data, and analytics to policy makers and other relevant stakeholders, CETI hopes to help craft better policy and accelerate action to meet ambitious, science-based greenhouse gas reduction targets. Eileen spends a lot of time considering how to actualize a clean energy transition that is beneficial to all, and particularly rural and under-resourced communities. This work is painstaking and challenging, but Eileen recognizes that the stakes are too high to fail. I so appreciate her tireless dedication to making strides in deep decarbonization one way or another.

Prior to founding CETI in 2018, Eileen accumulated decades of experience in nonprofit management and communication. She spent six years as a journalist covering business and national politics and another few years editing a Northwest quarterly public policy, economics, and culture journal. These experiences continue to profoundly shape how she communicates and strategizes about climate and energy to resonate with as many communities as possible. She also served as the executive director of several nonprofits before joining the Climate Solutions team as director of strategic innovation, where she worked with twenty-two cities and counties across the Northwest to identify pathways away from fossil fuels. While she enjoyed working in the weeds with cities, she felt like more support was needed at the state and regional levels to enact meaningful changes, and thus CETI was born.

Content Notes: Nuclear bombs, 9/11, death of loved ones

On the impetus of her climate action

My first foray into environmental activism was in the anti-nuclear movement in the late 1970s. My dad died of cancer when I was twelve, likely due to exposure to radiation; he was in the atomic bomb tests in Bikini Atoll.[1] But the turning point for me was really September 11, 2001. I'm from New York. I lost a cousin and a classmate in the towers and my family was deeply impacted. It was just a heartbreak on so many levels. Part of my analysis for why we were in that situation was our dependence on oil. As a country, we had been pretty hegemonic and terrible actors for a long time, particularly in the Middle East, and a lot of that was rooted in our need for oil. I became focused on our dependence on fossil fuels from a national security point of view, but very quickly I started to think more about global warming.

So, that was my awakening, but it took me seven years to get to the point where I said, *Okay I'm just going to work on climate for the rest of my life.* My dad only made it to forty-six and my mom had only made it to fifty-eight, so when I turned fifty, I was feeling the press of time. I decided to reassess. I looked ahead and thought if I get another fifteen or twenty years of work, this is what I want to work on. I wasn't initially sure how or where. Now I've been in this space for about a decade, and it's so complicated that there's no way I'll do anything else for the rest of my career.

On deep decarbonization and its resilience benefits

The five pillars of deep decarbonization are first and foremost:

1. Energy efficiency: using less energy throughout all processes, including in our buildings, industrial processes, cars, appliances—every end use possible

2. Decarbonizing the electric grid: further reducing the emissions intensity of electricity generation by shifting to clean energy sources

3. Electrification: since electricity is very efficient, if we're able to power transport and industrial processes with as much electricity as possible, then we can use the remaining carbon budget for the processes harder to decarbonize

4. Decarbonizing fuels: reducing the emissions intensity of liquid and gaseous fuels in the harder-to-decarbonize sectors like aviation and long-haul trucking

5. Sequestration: Sequestering the carbon emissions we cannot decarbonize.

Deep decarbonization has multiple major benefits for resilience. First, we reduce dependence on fossil fuels, which are still deeply problematic when it comes to national security concerns, but also because of their price volatility and their inherent danger. And not just the price volatility. An oil train can explode if it falls off the tracks or spills its contents. What happens if a coal train pours its coal into the Columbia River? There are enormous hazards involved with every aspect of fossil fuels. Extracting, burning, transporting, all of it. Less dependence on fossil fuels also means more reliability over time. We waste an enormous amount of energy when we are dependent on fossil fuels. They are extremely inefficient.

When I think of resilience, I also think about health. What comes out of a coal stack is arsenic and all kinds of really bad pollutants, beyond just the heat-trapping gases that cause global warming.[2] There are toxic chemicals that are released and break down when we burn fossil fuels, and they're hazardous for the water we drink and the air we breathe. The more that we stop that, the greater the health benefits, particularly for ailments like asthma.

I also think the clean energy transition could be a huge job creator. Having economic security is a resilience benefit. If we're truly going to address climate change, we will need to significantly reduce our consumption of goods that are shipped from really far away. If that means we build local economies up, that's a resilience benefit. Wresting control from multinational companies whose focus is profit, not people, is another piece of our opportunity here.

It would also mean empowering communities to be able to channel their own energy destiny. There are opportunities to do things differently that are threatening to fossil fuel companies and even to some utilities, but these are opportunities for people to safeguard the land and their health.

That being said, while we must invest in local, distributed energy resources, the honest truth is that we will still need the power grids that we have. If we really electrify as much as we must, Washington is still going to need to bring in wind from Wyoming and Montana and solar from Arizona, New Mexico, California, and Colorado. We're going to need to have a very robust clean energy grid throughout the eleven states in the West.

Local, distributed energy systems also tend to be more expensive, but not if you consider the co-benefits, and not if you consider the jobs, and not if you consider the resilience benefits. We are such a wealthy country, so it's really just a question of what we're trying to achieve. If we were to stop subsidizing oil and gas and coal companies tomorrow, and shift that into clean energy at the large utility scale, and also put a lot of money into community-led energy systems, we would have enormous benefits across a range of opportunities. Unfortunately, the most recent election in 2020 did not give us the green wave that would make that happen, so we're going to have to keep doing it more piecemeal.

It matters whether the Northwest achieves its decarbonization targets. We already have the cleanest electricity supply in the country. I believe that if we can't do it here, it can't be done anywhere. We really have got to prove it here. That gets me up in the morning to keep trying.

RESILIENCE TOOL SPOTLIGHT

Energy Efficiency

Energy efficiency forms the bedrock of decarbonization. It's an important first step in any decarbonization process because energy efficiency projects are relatively low-cost, straightforward, and less resource-intensive than large-scale energy infrastructure. Done well, they reduce the need for the latter. The energy efficiency umbrella is vast, generally referring to technological, material, and design changes that reduce energy consumption. The greatest opportunities for efficiency improvements lie in existing and new buildings, industrial processes, transportation, appliances, and equipment. LED light bulbs, building insulation, accessible public transit, and smart-home technology are all examples of mechanisms in the energy efficiency toolkit.

Energy efficiency may not be as buzzy as other forms of decarbonization innovation, but it's truly a resilience powerhouse with co-benefits galore. Like all the best resilience tools, energy efficiency efforts have the potential to advance climate mitigation, adaptation, and justice simultaneously. First of all, deep energy efficiency could halve the United States' energy consumption and greenhouse gas emissions by 2050.[3] In order to avoid the worst climate crisis scenarios and successfully transition away from fossil fuels, it will be imperative to dramatically reduce heat-trapping gases and the country's sky-high energy consumption. Secondly, community-wide energy efficiency efforts alleviate strain on the power grid, particularly amid extreme weather events when air-conditioning or heating systems may be working overtime. Energy efficiency can therefore help improve grid reliability when folks depend on power the most. Thirdly, energy efficiency is arguably one of the most effective water conservation strategies since traditional electric power generation is incredibly water intensive.[4] Power plants around the globe collectively consume multiple trillions of gallons of water each year.[5] As the climate crisis exacerbates widespread droughts and water scarcity, meaningfully reducing energy demand will help preserve precious water locally. Finally, energy efficiency programs create good jobs. The subsector currently employs more than twice as many workers as all fossil fuel industries combined and has the potential to employ many more.[6]

When looking specifically at weatherization programs, which seek to make buildings and homes more durable and efficient by adding insulation, sealing up holes in walls and windows, and updating appliances and equipment, there are even more resilience benefits:

- An effective weatherization program will result in lower utility bills. Older buildings often have inefficient appliances or poor insulation, and they often house lower-income residents, so upgrading those buildings can have meaningful economic benefits for those most impacted by structural oppression and poverty.
- A well-insulated, well-ventilated home can improve respiratory and cardiovascular health, reduce allergies, asthma, and headaches, and boost mental well-being.
- During a power outage, which often overlaps with heat waves, cold snaps, or major storms, a well-insulated home can retain heat or stay cool for a longer period, reducing serious health risks associated with extreme temperatures, particularly for elders and disabled folks. As extreme weather intensifies and becomes more frequent, weatherization will be instrumental in keeping folks safe.

It's important to note that energy efficiency alone does not guarantee a reduction in energy consumption, though. In fact, Jason Hickel warns in his book *Less Is More* that "when we innovate more efficient ways to use energy and resources, total consumption may briefly drop, but it quickly rebounds to an even higher rate . . . because companies use the savings to reinvest in ramping up more production" and because individuals and communities are less scrupulous with their energy consumption.[7] Particularly if the global economy remains tethered to the imperative of endless growth, energy demand will keep rising even if energy-intensive processes become more efficient. Therefore, to successfully transition away from fossil fuels, energy efficiency and other decarbonization efforts must be coupled with resistance toward the current growth-obsessed economic paradigm.

To get involved with energy efficiency efforts in your community:

- **Research prompts:** Which organizations or activists in your community are already engaged in energy efficiency work? What types of assistance or resources are they calling for? Which policies and incentive/technical assistance programs are already in place in your city, state, or region to expand and deepen energy efficiency efforts?
- **Journal prompts:** Jot down what a day in the life looks like for you. From cranking the air conditioner to laundering clothes to running errands, which behaviors demand the greatest energy consumption? What are the biggest structural barriers to "decarbonizing" your day? What would a deeply joyful day look like for you that requires very little energy consumption?
- **If you're passionate about energy efficiency and have the bandwidth:** Advocate for rigorous energy efficiency standards, policies, and programs in your city or state. Begin by tapping into energy efficiency advocacy efforts that are already underway in your community and looking to communities around the world who are leading the way. Study the legislation they've passed, and talk with organizers who have been involved with energy efficiency policy and program implementation efforts elsewhere. Keep in mind that rigorous energy efficiency standards can extend to new buildings, appliances, equipment, industrial practices, vehicles, and more, and advocating for energy efficiency can also look like advocating for more and better public transit, walkability, bike paths, affordable housing near city centers, and so on. Particularly if your community is rebuilding after a disaster, it's a critical time to be vocal for energy efficiency and resilience standards in new developments and planning.
- **If you're excited about energy efficiency but would rather avoid government spaces:** Be a consumer, employee, or investor activist.

Organize with others to demand that corporations that you shop at, work for, or invest in substantially reduce their energy use. Be wary of greenwashing; it's not enough for large corporations to buy carbon offsets or expand their renewable energy infrastructure if they're not simultaneously cutting their energy use. If you're engaged with a community institution, like a school, church, or recreation center, organize with others to weatherize buildings, prioritize sustainable design in any new buildings, transition the vehicle fleet to a more fuel efficient or electric option, and so forth.

- **If your home could benefit from weatherization:** Check out Benefits.gov to see if you're eligible for the Weatherization Assistance Program. Your city or state may also have some public incentives available that can be deployed for home retrofits, and your utility may have a Pay-As-You-Save program to help finance energy efficiency projects up front. If you live in an apartment building, consider organizing with other tenants to demand that your landlord weatherizes the building.
- **General best practices:** Note that while weatherization programs *can* have economic justice benefits, upgraded homes have historically been linked to rising housing costs, displacement, and gentrification. It's crucial to build relationships, listen to folks' concerns, and center the communities who would be most impacted by displacement to mitigate any unintended consequences of energy efficiency programs.

NATALIE HERNANDEZ

SHE/HER/HERS

"As an urban planner, my core role is really to be a liaison for the community"

Natalie Hernandez brings technical expertise and zeal to her role as director of climate planning and resilience at the Los Angeles nonprofit Climate Resolve. With a background in urban planning and environmental policy, Natalie has a wealth of knowledge about state funding, California's environmental programs, relationship building, and so much more. And as someone who has spent most of her life living in a community impacted by shipping and refinery pollution, Natalie's passion for environmental justice has been lifelong. She knows how to navigate complex institutions and bureaucracy while uplifting the concerns, desires, and solutions of people who will be most impacted by environmental and climate injustices. I was moved by the heart that Natalie brings to her role. Los Angeles will need to adapt its infrastructure in the years to come in order to keep its residents safe, and Natalie is motivated by that desire to protect her loved ones through extreme heat, sea-level rise, drought, wildfires and smoke, and other climate impacts expected to befall the area.

Natalie has coauthored California's Adaptation Planning Guide, conducted extensive climate-preparedness research, and led community outreach for resilience and adaptation projects. She also helped build Latinx Ambientalistas, a bilingual peer-support and speaker-series group created to connect Latine professionals and advocates working in the environmental space. In all functions, Natalie brings an understanding of how to lead meaningful transformation, from Los Angeles neighborhoods to siloed nonprofits and whitewashed environmentalism.

Content Notes: Environmental and climate injustice

On resilience across different scales and spheres

I think of climate resilience planning work as building blocks that can be stacked on top of each other. So for me, at the bottom there's the resilience plan for myself and my household: What would we do if there was a heat wave or earthquake, and how could we get in contact with my grandparents? Then there's the city of Long Beach, which is developing its own climate action and adaptation plan. On top of that, there's Los Angeles County, which has a sustainability plan and is developing a climate vulnerability assessment. And so on. I tend to work most often at community and regional scales, but every building block is crucial.

The technical definition of *resilience* is the capacity for a system to tolerate shocks or disaster events and recover and thrive. But a community can still be resilient even when there aren't shocks or disasters happening. In fact, it's important to be building resilience during the status quo. That's the time to have block parties, to get to know your neighbors, and identify where the emergency responders are. Building that social network is critical when times are normal. The more connection points that a community has, the more resilient it will be. We saw this play out recently during the pandemic. Neighbors who knew one another could check in and ask if they needed any groceries or additional support.

It's not just about neighbors helping neighbors, though. There are these different layers to resilience. Social resilience is crucial, but I also think a lot about physical resilience. If the power goes out during an extreme heat day, are there backup generators? Can the neighborhood power itself with its own solar panels and microgrid system? Are there plentiful jugs of water stored in case they lose access to water? The physical infrastructure piece is an important component of community resilience.

On strategic planning and design for climate adaptation

In the years to come, there's a lot that Los Angeles County will be affected by. We're going to see greater precipitation events and longer periods of drought.

Sea-level rise will impact some neighborhoods, and some areas will be impacted by wildfires. A lot of the greater Los Angeles region will be very affected by extreme heat. Temperatures are expected to rise about four to five degrees in the next century, and there will also be a greater frequency of extreme heat days, or days that are over 95°F. When extreme heat days occur, there are many people who don't go to work, and so their wages suffer, or they go to work and become dehydrated. There isn't an extreme heat day fund for people to dip into. My dad paints houses and my boyfriend is a sanitation worker, and I know there are days when it's really hard to work outside. I also think a lot about people waiting for transit when there's no bus shelter, just a bus stop sign (a very common sight in Los Angeles). My younger sister is in high school, and she's mentioned that some of her classmates miss the bus when they're standing in the shade to cool off and the bus doesn't see them. Many neighborhoods are already affected by air pollution, and they will be even more impacted by poor air quality as heat waves cause the air to stagnate and trap emitted pollutants. And since we're near the ocean, and much of the housing infrastructure is older, a lot of people will have to live with extreme heat without having air-conditioning or good ventilation.

Fortunately, there are so many urban planning tools that can improve environmental justice and climate resilience outcomes in our neighborhoods. Thoughtful and savvy zoning is one important piece. If you were laying out the blueprint of your house, you wouldn't put the toilet in the kitchen. Similarly, if you're laying out the blueprint of the city, the polluting industries and freeways should not be adjacent to residential areas and schools. It should be obvious, but that's not how so many cities are built. In fact, my mom taught kindergarten alongside the truck corridor transporting goods from the port of Long Beach to the rest of the country. There were also refineries nearby. My mom had to advocate for better air quality for her kindergarteners, who were largely from low-income, immigrant families, because they were already dealing with asthma and respiratory diseases. Prudent zoning can help avoid or ameliorate these environmental injustices.

There are also several urban design techniques that can be used to increase a neighborhood's capacity for resilience. For instance, I've been working on an urban cooling project in Canoga Park, a neighborhood in San Fernando Valley, where they're used to getting extreme heat days over 100°F. Trees are obviously great, but they're hard to plant everywhere, unfortunately. Cool roofs, which can be made with a highly reflective type of paint, reflect more sunlight and absorb less heat, so the buildings themselves stay cooler. Cool pavement, which is lighter in color than traditional asphalt and uses additives to reflect solar radiation, can be utilized on residential streets. Residents in Canoga Park have said that they feel a five-degree difference between streets with cool pavement and those without. Shade structures for bus stations are important, as well as hydration stations where people can fill up their water bottles.

I'm working from the understanding that Black people, Indigenous people, and people of color, low-income folks, and immigrants bear the brunt of climate change. Therefore, advocating for policies that are going to help those communities, like affordable housing and equitable transportation, is also super important. And it's a central piece of climate change work. Connecting the dots between interrelated issues is pretty simple in theory, but I think we've overcomplicated it. We have these conversations over and over again, explaining how housing connects to transit, which connects to wildlife preservation, which connects to climate change. Thanks to social justice advocates, more and more people are starting to get it.

As an urban planner, my core role is really to be a liaison for the community. The residents are really the experts of the things going on in their community and which solutions will work best. Therefore, my team and I spend much of our time talking with community members and working to deeply understand their perspectives. We work hard to meet community members where they naturally gather and tailor our outreach to them. We hold workshops at the community farmers market that folks frequent and host presentations at the local high schools. We conduct surveys at transit stations. We translate wonky or confusing data so that folks can understand what impacts may mean for them. All our materials are bilingual. I can speak Spanish, as can some of my colleagues, so we can speak with ease to predominantly Latino community members. In every instance, we're trying to meet the community where they're at and relate the issues to what they're concerned about. Our primary goal is to uplift the expertise in the community and get knowledge to the people so that they can take action themselves.

RESILIENCE TOOL SPOTLIGHT

Planning and Design for Urban Cooling

As heat waves become longer, hotter, and more frequent, urban communities experience the impacts most intensely; cities can be up to 22°F hotter than surrounding rural and suburban communities.[1] The urban heat island effect results from large swaths of concrete and asphalt, which absorb and radiate the sun's rays, as well as urban canyons formed between tall buildings that trap heat at the street level. Not every urban neighborhood is affected equally, though. There is tremendous temperature variation within cities, typically mapping with the racial and socioeconomic history of the region. On average, communities that have been structurally marginalized have significantly fewer trees and green spaces and more parking lots and pavement. In fact, a 2020 study of 108 urban areas across the United States found that formerly redlined neighborhoods of nearly every city studied were notably hotter than nonredlined neighborhoods, sometimes by nearly 20°F.[2] Low-income communities and communities of color are therefore disproportionately impacted by the urban heat island effect.

There is, however, an abundance of design and planning strategies to help keep communities cool and safe during heat waves. Plants in all forms, including green walls and roofs, parks, community gardens, and tree plantings, can significantly reduce temperatures. A climate researcher recently found that walls covered in vining plants had surface temperatures 40°F cooler than their bare-walled counterparts next door![3] Creating safe, appealing alternatives to car transit, like walking, biking, and taking the public bus or train, can help take heat-producing cars off the street and shrink the size of heat-absorbing roads and parking lots. Simple shade structures and water features, like fountains, sprinklers, and misting systems, can give folks a cool respite at bus stops and in public gathering spaces. Light-colored and reflective coatings applied to roads, roofs, and building facades have proven effective in reflecting rather than absorbing heat. Black asphalt roads in Los Angeles were shown to be a whopping 23°F cooler after they were coated with a white, highly reflective sealant![4]

No single strategy is a panacea. In fact, if these urban cooling tactics aren't utilized with tremendous thoughtfulness and community engagement, they likely won't be particularly impactful, and may even end up displacing community members. Cooling solutions should always be community-led, as local residents will always be the foremost experts on where it's hottest and which solutions will make sense for the community and why.

To get involved with urban cooling efforts in your community:

- **Research prompts:** How is your community expected to be impacted by extreme heat in the years ahead? How might extreme heat be experienced differently across your community, and how has that variation been shaped by historical policy and design choices? Which local organizations and organizers are working to ameliorate the urban heat island effect and related issues? Which public spaces could provide cool respite during a heat wave?
- **Journal prompt:** Take a moment to dream: Where might you feel safest, coolest, and happiest on an extremely hot day? What does this place look, smell, taste, and feel like? How could you incorporate some of those elements into your home and community?
- **If you have access to time and a passion for data:** Be a citizen scientist. Whether on your own or with community members, attach a thermal imager to your smartphone and begin collecting data. As you move around your neighborhood, make note of temperature extremes and any patterns that arise. Particularly if you're able to gather a group of folks to acquire data in different areas of the neighborhood, this information could be invaluable in helping guide local government's planning and heat wave response efforts. The same exercise can be conducted by getting in touch with your senses and making note

of locations that feel the warmest and coolest, particularly on a hot, sunny day.

- **If you have access to time and shareable skills:** Lend your talents to organizations and organizers who are working to shift policy, planning, design, and emergency response to keep all community members safe amid extreme weather. Help amplify their work, fundraise, share your own expertise or resources, show up when volunteers are needed, recruit friends and neighbors to show up at town halls and community gatherings, and galvanize folks to submit public comments.
- **If you have flexibility during your day and/or access to transportation:** Help care for the community members most vulnerable to heat wave impacts by coordinating rides to cooling shelters and shaded parks, making phone calls or in-person check-ins to neighbors, or distributing care packages filled with items like chilled water bottles, sports drinks, ice packs, snacks, towels, hats, or umbrellas to unhoused neighbors.
- **If you have access to a robust community network:** Gather neighbors to discuss cooling strategies that could be utilized on your block. Identifying favorite plants and agreeing upon a few places to add vegetation (such as the hell strip in between the sidewalk and street curb) are great places to start. Note that your city may have a specific permitting process or programs in place to support street cooling efforts, so volunteer or appoint a neighbor with the bandwidth to check local ordinances, liaise with the city, and coordinate logistics.
- **If your workplace is impacted by extreme temperatures:** Organize coworkers to discuss and demand weather-related boundaries. If you are in a position of power at work, take the lead in listening to what your coworkers need and implementing inclusive policies designed to prioritize worker safety and health through extreme weather events.
- **If you have a green thumb:** Bring more plants into your home. Just as vegetation brings temperature relief to the outdoors, indoor plants

cool the surrounding environment when they release excess water into the air from their leaves. Prioritize plants with lots of foliage, large leaves, and varieties that thrive in humid environments.

- **General best practices:** Whether advocating for policy changes, supporting an organization's ongoing efforts, or helping develop a cooling project in your neighborhood, make sure to center the folks who are most impacted by extreme heat and the urban heat island effect every step of the way.

MINDY BLANK

SHE/HER/HERS

"My work focuses on creating a bridge between collapse and a more just future"

Mindy Blank is the executive director of Community Resilience Organizations, a nonprofit that emerged after the heavy rains of Tropical Storm Irene devastated communities across Vermont in 2011. The organization was founded to help Vermont's towns and cities strengthen their resilience to climate change, economic volatility, and other unforeseen disruptions by relocalizing critical resources like food, water, and energy, strengthening social connectivity, and tapping into place-based strengths. As a team of one, Mindy embodies myriad roles to bring this mission to life. On any given day, she might be co-designing a new community resilience program with librarians, teachers, and farmers, distributing grants, conducting a community assessment, or facilitating a collective grieving circle. Her work is deeply heartful and collaborative.

As a child of immigrants who has been a part of numerous communities, from a small island town on the Mississippi River to the cloud rainforests in Panama, Mindy is seasoned at cultivating meaningful relationships with people and places wherever she goes. She is also particularly tuned in to the communities that have been and will be displaced by the climate crisis, and knows that Vermont, dubbed a "climate haven" by researchers and media pundits, will likely receive many climate migrants in the coming years. I sought out Mindy's perspective in particular because too few communities are considering how to proactively prepare for climate migration in ways that honor the needs and dignity of their current residents and the humans that have been displaced from their homes, but Mindy isn't shying away from the challenge. Instead, she's making space for much needed community-wide conversations with tremendous nuance and care.

Content Notes: Allusion to interpersonal xenophobia and racism

On doing the work that's meant for you

I thought that working at the International Energy Agency in Paris would be my dream job. I was tasked with creating roadmaps for governments to transition away from fossil fuels in a just way, and it felt big. I had always wanted to be working at the international scale, because it felt important that I was using my capacities in the exact right space to impact the most change. I really thought that was what it was all about. And then I remember working with a United Nations Environment Programme group on a renewable energy project and having this moment of realization that I actually didn't think I could get behind the project with my full heart because it was still extractive. It wasn't as continually extractive as fossil fuels, but it was still going to decimate communities. So many of these renewable energy projects are still decimating our land and water.

So I returned to Vermont and decided to focus my energy at the local and regional scale as a community organizer. Now I get to use my skills and passions to try to answer the question, *How do we help communities across the state better prepare for disasters, and how do we create better community relationships to be able to do that?* In essence, my work focuses on creating a bridge between collapse and a more just future. It's about learning from the past, actively addressing pressing realities of the present, and doing so with the future in mind. It's systemic work and it's deep, personal work. It's untangling white supremacy culture. It's acknowledging the magnitude of the complexity of these crises. It's holding incredible grief as well as joy and beauty. It's all of this together.

On finding hope in agriculture

One of the places that I feel the most hopeful is in the food system space. Vermont is an agricultural state, and the farming community is among the most impacted by climate change here. There are a lot of real concerns, and a lot of damage has already been done. At any given point in the summer, some area of Vermont will be in drought and another area will be experiencing downpours and floods. It's becoming increasingly hard to plan. The land and water systems are very vulnerable to the changing climate.

At the same time, farmers are so knowledgeable and they're naturally movement leaders. Farming is a place where transitions can really happen. During the pandemic, we saw the capacity of our regional food system rapidly increase. The farm that I live on went from providing 100 CSA shares to 300 shares in the past year. That's a common story across the state. The food system web grew stronger. So when we talk about building a bridge from collapse to a more just future, farming is a piece of that bridge. When we relocalize our food systems, we no longer need to rely on the large-scale, industrial, capitalist systems to meet our basic human needs. We can actually step away from those oppressive systems. That work is already happening and needs to happen faster.

On shifting culture and making room for climate refugees

Vermont's biggest climate-related concern is probably the influx of new people coming here. From a climate change perspective, this will be one of the safer places to live now and into the future. People will keep moving here. So I believe that one of the most important pieces of climate resilience work in Vermont is cultivating communities that are truly welcoming to climate refugees and other folks coming here. Vermont is such a dominantly white state (94.2 percent white, per 2021 US Census Bureau data). If there isn't a social and cultural shift now, I fear for the violent confrontations that will occur when climate refugees inevitably flock here. This work is essential immediately.

I sometimes grapple with doing this work in rural Vermont. It's definitely not the place that is being hit the hardest by climate change, so why am I not in the places that need more support? This work is personal to me, though. Because I've moved a lot and I'm a first-generation American raised by immigrants, I haven't ever necessarily really felt home in a place. I didn't really grow up with a sense of home. It's important for me to be part of a movement to create a sense of home for all that can exist in the future.

On defining success

We're often asked how we measure this work. How do we see that it's working? When you take a look at the resilience projects happening across Vermont, you get this amazing, big picture of resilience; communities are doing everything from planting riparian buffers to building emergency shelters to creating community bread-baking ovens to repurposing parks as common spaces and making buildings more accessible. We use a community resilience assessment to gauge progress, and we have some measurable data on those projects. But the work is quite spread out, and none of it is enough.

I don't have the larger indicators for climate resilience success, and I don't know if I would believe them if I did. So we can talk about success in a different way. I like to look at it as these little shifts in the social dial. With each conversation, event, or interaction, something shifts. It's incremental. I see it in relationships with people. In mutual aid. In the development of community resources. I see our webs of relationships growing in stronger ways. It's these culture shifting pieces that feel like the real moments of success for me.

On how to get started organizing

Bring pie and make friends! A huge part of this work is just showing up in other people's spaces and forming relationships, not with a specific agenda other than getting to know them as humans. If you want people to be a part of your thing, you can't just expect that they're going to have any interest or time. But if you show up for their things and bring something to the table, sometimes literally, I think that makes a huge difference in this work.

CHIEF SHIRELL PARFAIT-DARDAR

SHE/HER/HERS

"Our home is our identity"

As a minister, musician, traditional dressmaker, business owner, mother of four, avid gardener, environmental advocate, and the first woman traditional tribal chief of the Grand Caillou/Dulac Band of Biloxi-Chitimacha-Choctaw, Shirell Parfait-Dardar fills myriad roles that appear disparate in nature. Upon closer inspection, however, it's clear that everything Shirell does is deeply rooted in and driven by deep love and care for her ancestors and the next seven generations. In each capacity, she is a steward of her people, their culture, their more-than-human relatives, and the land and water that they are a part of.

After witnessing the bayou communities of coastal so-called Louisiana erode in the rising waters and threaten to be entirely consumed by the Gulf with each passing hurricane, Shirell knew that her tribe was at risk of being among the world's first climate refugees. She felt compelled to step into the chief position to carry the voices of her people into rooms where they needed to be represented and to lead policy change. Upon recognizing that other tribes were experiencing similar issues, Shirell cocreated the First Peoples' Conservation Council of Louisiana, an association of six tribes dedicated to protecting and restoring their lands, waters, and air. As the chairwoman of the Louisiana Native American Commission and as part of a formal complaint to the United Nations citing that the United States government failed to protect the human rights of five tribes in so-called Louisiana and Alaska, Shirell has earned the name Killer Red Fox because she's known for attacking and killing bad policy and for protecting her kin. Climate-induced community relocation is extremely complex, and there's no roadmap for how it should be done. It was an honor to speak with someone who is fighting so fiercely to get it right for her people.

Content Notes: Settler colonization, anti-Indigenous discrimination, forced displacement

On the changing environment and threat of relocation

I grew up in what's known as Grand Caillou Dulac along the predominantly Indigenous area known as Shrimpers Row. I currently live in Chauvin, which is the next bayou over. Over the decades, we've seen our landscape change drastically, and quite quickly.

At one point, when we were children, there were tons of wooded areas where we could play and feel safe. Now when you drive down into our historic tribal communities, you look out and can see all the trees that have died from saltwater intrusion and erosion.[1] Where there were once thriving, beautiful landscapes, there are just little slivers of land left. You look out and can see death coming.

We are a water and land people. We rely on the shrimping and crabbing for our sustenance and for our income. With all the environmental changes and land lost to sea-level rise, we're losing those habitats. The shrimp and crabs have lost their breeding grounds, and that drastically impacts our ability to rely upon those fisheries. We can no longer draw an income from traditional practices. We can only do it as a means to feed ourselves, and even that is becoming difficult. It's heartbreaking, because historically food has been everything for us.

Meanwhile, we're facing forced displacement. With every major storm that hits Shrimpers Row, people are forced to move to higher ground. Because of storm and flood damages, insurance rates are now astronomical. People can't afford to insure their homes. It's breaking apart our tribal community, which is detrimental for a few reasons. First of all, as we pursue federal acknowledgment, it's critical that 50 percent of our tribal membership resides within our community.[2] That's becoming harder and harder to do. Second of all, with our people being forced to leave our community and assimilate into new communities, we lose familial connections and connections to place.

We're currently looking into resettlement, which is extremely scary. We saw what happened with the Isle de Jean Charles community when they tried to work with the government to relocate their community.[3] We're hoping to find alternative pathways to relocation so that we can retain our culture and our independence as Indigenous people, but just trying to keep our people together is such a challenge.

It's also very hard to find suitable land in Louisiana. So much of the land is owned by oil and gas companies or different land developers, and the price tags on that land are extremely high. Even still, the large areas of land are so far inland that it would literally remove my people from who they are. It's traumatic to have to take a person away from the home that they've built and known for many decades. Our home is our identity, literally.

And since there aren't any programs in place to transition our people into new economic opportunities, forced displacement is pouring salt onto an open wound. Our entire livelihoods have been dependent upon traditional practices. Now we are expected to leave our homelands and move into a foreign, inland area to survive, while we're stripped of everything that we've been doing our whole lives for economic stability.

For me, I moved to Chauvin thirteen years ago, and it took years to adjust to the fact that I'm off my bayou, even though I'm still within my tribal land. I was so mad that it took me thirty minutes to get back to Shrimpers Row. But the ground is higher here. There are stores here that we don't have in Dulac anymore. There's an opportunity for jobs for my kids here that we don't have in Dulac anymore. My family made a decision based on opportunity and safety. I've adapted to being here, but it took me a very long time. So I've personally experienced relocation on a very small scale. I don't even want to imagine what my people will go through if I must move them an hour away from where they're used to being, in a whole other community, surrounded by strangers that they do not know. Trying to figure out how to navigate that has been gut-wrenching, to say the least.

On success and leadership

I don't know if success can be measured in this situation, but I ask myself, Am I making progress? Is what I'm doing beneficial to all, not just a contained group? Are we negatively impacting anyone? And as long as we've been able to make progress, no matter how big or how small, progress is progress. It brings me peace when I'm able to see that we've done something to protect and preserve our future generations and that our tribe is going to continue. And I'm hopeful that, one day, our youth will take the reins and carry us even further.

Without being able to see that the work that I'm doing is going to contribute to future generations, I don't think I'd be able to do it at all. While I don't get attacked as much as I used to when I first started the work, thanks to civil rights activists who continue to shift culture, it's still a challenge. Just being an Indigenous person is a challenge. Even today.

I've often told people, "I'm going to kick ass but I'm going to cry first." It's okay to cry after going through some of the things that we go through. But it's not going to stop me. My duty is to represent my people to the best of my ability, and the best way to do that is to carry *their* voice, not mine.

Cultural Strategy

While the climate crisis represents both a literal and a figurative collapse of present worlds, it also serves as an invitation to dream up and actualize new worlds. These essays are from artists, cultural activists, healers, and storytellers who are nurturing imagination, expanding conceptions of what's possible, and uplifting the ideas and beliefs needed to build a just future. They are committed to embodying the transformation that they'd like to see and bringing much-needed soulfulness, artistry, and play to the climate movement.

EVE MOSHER

SHE/HER/HERS

"Artists are really good at showing what's wrong with the world—it's time to show what's possible"

Eve Mosher is a globally recognized artist, interventionist, people-centered researcher, and parent. Eve began her career as a studio artist, interested in exploring the intersection of natural and built environments through drawings, sculptures, and installations. But after growing frustrated with the discourse and inaction around the climate crisis, she decided to pivot into the public realm to engage a larger audience. HighWaterLine NYC was born. In 2007 Eve took to the streets of Brooklyn and Lower Manhattan with copious amounts of scientific research on projected sea-level rise and flooding and a Heavy Hitter, a machine typically used to draw chalk lines on ball fields. She walked seventy miles around New York City, drawing blue chalk lines on the pavement to demonstrate where climate change would create more frequent floods. While walking, she struck up conversations with the folks who lived, worked, and played in those spaces, listening to their stories, sharing educational materials, inviting them to community workshops, and welcoming them to consider how to respond.

Over time Eve's work shifted from dialogic to deeply collaborative. Today her role is primarily that of facilitator, and she specializes in developing the infrastructure to spark radical imagination and sustain creativity to help communities tell their own story of tomorrow. She has partnered with countless individuals and organizations to create HighWaterLine projects around the world. She has also cocreated countless immersive multimedia experiences and works of public art around the globe, all of which are intended to foster connection, inspire climate solutions, and produce visions of possible futures. It's a joy to uplift Eve's work and the movements that she's nurturing, as well as spotlight the power of underutilized climate action tools like art, play, and radical imagination.

On her journey into environmental artwork

I grew up just outside of Houston, Texas, at the mouth of the Buffalo Bayou. We were on the edges of Texas wilderness and development and watched the urban sprawl unfold around us. That massively informed a lot of who I am today. I went on to study architecture and then art, always focusing on this intersection of humans, lived experience, and the built environment. A lot of my work was trying to tie nature into the urban environment, which was sort of the opposite of what I saw growing up, as urban sprawl encroached upon the wilderness.

As my work as an artist evolved, I was very influenced by the Bush administration's denial of climate change, and opposite that, this constant drum beat about climate change as it related to polar bears. It was all being framed as very distant, in both time and space. Of course, there are humans who live alongside polar bears, and that's an important piece of the climate change conversation, too, but my parts of the world happen to be urban areas, and those places weren't being talked about. That's what got me to take my work out into the streets, literally.

On HighWaterLine and the importance of community collaboration

In 2007 I began the HighWaterLine project in New York City. HighWaterLine, at its core, was about this drawing of a chalk line that showed people which areas of their community would be susceptible to increased flooding and storm surge because of rising and warming seas. I was inspired to do this because I see coastal communities as being both under threat and at a point where there's great opportunity. There's an opportunity to learn from communities around the world who already know how to live with water, and also to shift our perspective. My friends Dilip Da Cunha and Anu Mathur like to say, "Stop thinking of colonizing the water. Get away from the idea that water has hard boundaries, and learn to accept degrees of wetness." I think that coastal communities have space to learn from many other communities around the world and have space to be innovative in thinking about how we live in a changing world.

The chalk lines didn't last, but the point was really to start conversations. It was about creating something that was intriguing enough that people would come up and talk to you. We had local climate researchers out there, and they understood and could convey the science, but they weren't there to lecture. It was really about exchanging personal stories and connecting the project to lived experiences.

The project eventually expanded to Miami, Philadelphia, and Bristol. HighWaterLine NYC was definitely a predetermined project (with consent from and engagement with community members), but as I had countless conversations with the people who lived in the neighborhoods I was walking through, I started recognizing how much knowledge and creativity already exists. I no longer wanted my work to have a predetermined outcome, and instead I wanted to be able to provide tools and resources to facilitate space for conversation and creativity.

I also started collaborating with Heidi Quante, who has tremendous community engagement experience and who I've learned so much from. (Heidi is an interdisciplinary artist who has been engaging the public through art-based environmental and human rights initiatives since 2003.) With her help, we really shifted the focus to building communities of collaboration. In Miami, for instance, the HighWaterLine project went through areas that were massively diverse geographically. We'd have an artist mom located in a northern suburban neighborhood collaborating with an architect based in Miami Beach and they'd be sharing resources and knowledge. The more that we're able to connect people to one another, the more creative our problem solving will become. The tighter that we're able to weave this web, the greater our resilience will be.

On current and future projects

When this work began about fourteen years ago, the focus was on both the problem and the solutions. At this point, we've rung the emergency bell long enough and it's time to focus on the solutions. Artists are really good at showing what's wrong with the world. It's time to show what's possible. We know what we don't want. But what do we want instead? All my work today is about creating a figurative space for exploring those ideas, having conversations around those ideas, and taking action on those ideas. All my projects are meant to create space for that radical imagination.

Right now I'm working on two overlapping projects in Philadelphia: Heat Response and Story of Tomorrow. I'm cocreating Heat Response with the Trust for Public Land, an organization that works really hard to create green space in urban environments. I love that their tagline is *Parks for People* because I never want to absent people from the conversation. We're codeveloping a framework for local artists, community leaders, and city agency folks to come together to dream up creative ways to address extreme heat, which Philadelphia is already suffering tremendously from.[1]

The Story of Tomorrow will be an immersive audiovisual story of tomorrow. The basic premise of the project is that another world is possible. We're asking, *What does that world look like, how can we elevate that in a way that allows other people to see themselves in it, and how can we scale up the solutions that are already here in this community?* It will walk community members through imagining what their neighborhood is like if we do everything right. The experience will include projection mapping and intimate, in-ear audio stories that connect the present and the future. I like that it's something people can experience together but separately.

I'm excited to bring the Story of Tomorrow to a lot of different places, tailoring the project to the unique issues that the city is facing and collaborating with the groups and activist networks that are already doing on-the-ground work. We want the Story of Tomorrow to inspire folks to see what they can do and then get busy doing it, so it's important that we're able to connect them with folks who are already doing that work. Heidi taught me that you don't just do a project and leave; you run a hundred yards past the finish line. We have to be intentional about maintaining a solutions focus and weaving connections together, so that there's continued growth, focus, and impact on these issues once we eventually leave. For me, success lies in that ripple effect.

My daughter and I were recently planting our seeds for the spring, and we were just so fascinated by these tiny little seeds that will eventually turn into these big plants. That's sort of how I feel about all these projects happening around the world right now. There are all these seeds that are just bursting with potential. They are growing and will continue to grow.

IRFANA JETHA NOORANI

SHE/HER/HERS

"What emerges when communities create together can't be imagined in siloed spaces"

Green spaces and community gathering hubs are both foundational to climate resilience, but when they're introduced into a community without meaningful community involvement and care, they can trigger displacement and other forms of harm to existing residents. So once I learned about Irfana Jetha Noorani's incredible work with the 11th Street Bridge Park, a collaborative effort between the District of Columbia and local nonprofit Building Bridges Across the River to repurpose an aging bridge into a seven-acre park, I knew her insight needed to be included in this book. The park will connect the Washington Navy Yard, considered to be one of the most gentrified neighborhoods in Washington, DC,[1] and the historic district of Anacostia, a predominantly Black community that has experienced economic and environmental oppression.[2] During Irfana's six-year tenure as founding member and deputy director of the park, she helped shape a cutting-edge, resident-led, and equity-first model for civic space planning and regenerative community development. She collaborated closely with community members to develop strategies to resist gentrification and displacement and preserve and enhance local arts and culture. Irfana also raised tens of millions of dollars to pour into small businesses, affordable housing, job training, and cultural programming in surrounding neighborhoods. Upon completion, the 11th Street Bridge Park will include an environmental education center, playgrounds, community gardens, performance spaces, public art, walking and biking paths, and more.

In 2020 Irfana transitioned from deputy director of 11th Street Bridge Park into a senior consulting role, focusing more specifically on resident engagement and collaborative public art planning. She has since been able to bring her deep expertise in equitable planning processes and programming to other adaptive reuse projects around the country, helping convert aging infrastructure into major green spaces in the most just way possible. She serves as a senior fellow to the High Line Network, promoting peer-to-peer exchange and partnership among a group of infrastructure reuse projects across Turtle Island. When Irfana isn't helping actualize community-centered culture and civic planning projects, she is dancing, disrupting white supremacy in philanthropy, and nurturing artistic community.

Content Notes: Structural racism, environmental injustice

On the evolution of the 11th Street Bridge Park project

When I started work with the 11th Street Bridge Park as a founding staff member in 2014, it was very much a park design project. There had already been deep community engagement, and the community was working with us to design and drive all the assets that were going to be included in the park. But the project was really centered on the space. As we went through a design competition and continued to have an enormous amount of conversations with residents, we listened really deeply. In our discussions with community members, we heard the need for jobs, the hesitation around gentrification, the desire for support for small businesses and in particular Black businesses, and the lack of access to capital for those businesses.

At the same time, we were looking at projects happening around the country, like the High Line in New York City, the 606 in Chicago, and numerous projects in Texas.[3] The data told us that there would be displacement unless we intentionally made sure that we focused on equity and inclusivity. We asked ourselves, *What can we learn from these projects that have already been done, and how can we do things better and more equitably?* Since we're located in the nation's capital, we have tremendous access to people who do this type of work nationally and who see what's happening across the country, and also who care about the projects happening in their neighborhoods. We were able to call those experts in and bring them together to think with us about how we could change development patterns in our neighborhood.

In 2015 we developed the Equitable Development Plan with our community, and the whole project flipped on its head. We decided that we would start with the neighborhood and the infrastructure would come later. This concept of *placekeeping* became central to the project. We frequently approach these developments by looking at economic metrics, like housing, small businesses, neighborhood development, and job training. Very often we forget about the feeling of being in your neighborhood, the feeling of belonging, and how spaces and people can change to either make you feel welcome in your neighborhood or not. It was really important to me that once we figured out the economic pieces, we figured out how to address the cultural pieces and ensure that we were keeping people at the center of our work.

The cultural equity and equitable development strategies came straight from the residents and artists living in the neighborhood. We went in with no idea what was going to come out of it. Our work was driven by the question, How can we use 11th Street Bridge Park as a platform to amplify the work and voices of Black residents and Black artists in the neighborhood? How can we gather resources and then turn it over to artists to do what they do best?

There are so many different placekeeping models, whether supporting artist-led projects or doing partnership-based work. I was with the 11th Street Bridge Park for six years, and we're just figuring out what this looks like in practice. There's still so much room to grow.

On the expansive possibilities of community gathering spaces

Now that I've worked in the public park space, I'm just fascinated by how critical it is to civic infrastructure and community building. In this pandemic era, it has become really apparent how critical our neighborhood spaces are in seeing each other. I'm actually super pumped about how public space has been the first place where we can come back together. It speaks to how critical this type of infrastructure is to our neighborhoods.

With 11th Street Bridge Park, we're building infrastructure to connect two communities together. It is going to be really important for the growth of our city and for people who don't often see each other to share space and experiences together. It's really an opportunity to consider equity, inclusion, and power sharing for these neighborhoods, and that is really helping us think about how we can responsibly develop in these neighborhoods moving forward.

I'm also hoping the park helps us deal with climate resilience in the future. Of course green infrastructure will be crucial. In particular, though, I'm excited about our programming. I look at it as a way to make the space come to life and to have our community members lead environmental and climate change programming in a way that really speaks to what residents are experiencing.

On the arts as essential to community development

After working in the community development space for the past six years or so, it's been striking to see how often the arts get left out of the conversation and how absent the arts are from community development spaces altogether. People don't think about how critical the arts are for healing, for empathy, for resilience. When you have amazing people and artists around you, they allow you to work from an asset-based place rather than a deficit-based place.

They also remind you that our efforts are better together. In all these places I'm creating and spaces that I'm working in, the culture of partnership and collaboration is just so pervasive. There's a culture that things cannot be done by yourself. People forget what comes from communities when they create together. What emerges can't be imagined in siloed spaces. The pie is bigger when we all build it together.

RESILIENCE TOOL SPOTLIGHT
Community Gathering Spaces

The presence of well-loved community gathering spaces, such as parks, houses of worship, local cafés and diners, and libraries, is an absolutely critical element of climate resilience. When working to better understand the outcomes of the devastating 1995 Chicago heat wave,[4] which killed over 700 people, public sociologist Eric Klinenberg found that close connections to other community members made people far more likely to survive. After he poured over quantitative data and couldn't find any explanations for the disparities in heat-related deaths between neighborhoods, he took to the streets and found that the communities with lively foot traffic, well-maintained parks, and an abundance of spots for neighbors to convene had fared much better. Meanwhile, in both extremely affluent neighborhoods and neighborhoods impacted by structural racism, poverty, and disinvestment, neighbors that were more isolated from one another, particularly elders and disabled folks living alone, fared worse.

Christine E. Nieves Rodriguez explained in her essay in *All We Can Save*, "When everything collapses, our life-saving infrastructure is our knowledge of one another's skills, our trust of one another, our ability to work with our neighbors, and mobilize. You have to be ready to tap into a pre-existing system of trust."[5] Therefore, it is a mistake for climate adaptation and resilience efforts to focus solely on physical infrastructure while neglecting the spaces that foster connection, collaboration, and support. It's also important to note that communities shouldn't have to choose between one or another; social infrastructure designed thoughtfully can have a multitude of co-benefits. Large parks and gathering spaces built by rivers or the coast can double as water basins or help soak up storm surge. Libraries can be retrofitted to generate and store renewable energy so that folks can access air-conditioning, heating, power, and other forms of support during a power outage. Houses of worship can prepare to utilize their industrial kitchens, childcare rooms, and gathering places to feed the community, mobilize volunteers and resources, and provide care during and after disasters. The list goes on.

That being said, gathering spaces that work well for one community may not work for another, and new spaces may even trigger gentrification and other forms of harm. Therefore, it is crucial that new projects are truly for the community and by the community. Irfana's work with the 11th

Street Bridge Park provides a fantastic blueprint for what placekeeping and a meaningfully community-driven process can look like in practice. Klinenberg also notes that most communities would benefit from fewer massive, high-cost initiatives and more investments in existing infrastructure, small-scale parks, accessible community gardens, and the like.

To get involved with the movement to invest in your community's gathering spaces:

- **Journal prompts:** Where do you feel most connected to your community or communities? Where do you feel the safest? The most supported? Now take a moment to envision what doesn't already exist. What might your ideal community gathering space entail? What sensations do you experience when you're there? What does it look, sound, smell, taste, feel like?
- **If you're engaged in a faith community:** Strategize and prepare with members of your house of worship so that your synagogue, mosque, temple, or church can serve as a hub of support for the local community before, during, and after climate-related disasters. Consider how you might tap into the house of worship's facilities, congregation, and other assets to provide hot meals, childcare, counseling, bodywork, a place for folks to sleep, a site for supply storage and/or distribution, and so on. Might it be possible to equip the building with renewable energy generation or storage to provide heating, cooling, and power amid a power outage? To get started, check out organizations like Creation Justice Ministries,[6] which has a robust climate resilience initiative, and Communities Responding to Extreme Weather (CREW), which hosts an annual Interfaith Summit.
- **If you're a lover of libraries:** Support continued investment in your local library by signing up for a membership card or dusting off the one you already have, joining your local Friends of the Library chapter, contacting your representatives to share the value of the library in your life, attending programming and events, and shopping at their book sales.

If you work or volunteer at the library, explore the American Library Association's Resilient Communities Programming Guide to weave more climate education into film and book clubs, community dialogues, and afterschool offerings. Explore the Sustainable Libraries Initiative for resources on transforming your library into a climate resilience hub. Consider what retrofits, updates, and additional investments would be needed for your library to continue offering community services during and after the disasters most likely to impact your neighborhood.

- **If you're deeply engaged in another community space:** Team up with others to consider how you might elevate the space to become a hub for resilience and mutual aid. Think outside the box: art studios, urban farms, community centers, cafés, and brick-and-mortar stores can all be equipped to meet various community needs during and after disasters—and to serve as a gathering spot for folks to connect, strategize, and build people power year-round. Check out free resources like *The Resilience We Want* guide, available as a PDF at Shareable.net, and the *Guide to Developing Resilience Hubs,* created by the Urban Sustainability Directors Network and available as a PDF at www.resilience-hub.org, to get started. Think big and start small.
- **If you work in city planning, architecture, development, or design:** Advocate for the development of safe and accessible community spaces where all kinds of people can gather, or invest in existing spaces that already serve this purpose. Are there any places that are currently being underutilized, like vacant lots or parking spaces, that might instead enhance the physical, emotional, and social well-being of the neighborhood? Strive to maximize climate mitigation, adaptation, and justice co-benefits as much as possible. Conduct deep and meaningful community engagement so that current residents, and particularly those most impacted by structural oppression, are in the driver's seat of the project.

MELISSA REYES

SHE/THEY

"Storytelling is about reclamation, right relationship, and healing"

Melissa Reyes is a healer. As a full-spectrum birth worker, clinical herbalist, ceremonialist, bodywork practitioner, and aspiring death worker, she finds guidance from her ancestors, who descend from Borinquen and Poland, the teachings of plants, and the folk healing system of curanderismo. Her practice, Remedios Birth & Healing, specializes in loving, accessible, trauma-informed, and culturally appropriate care and support for queer, trans, and BIPOC folks. She is deeply committed to reclaiming ancestral wisdom and Indigenous ways to help address generational trauma, cultivate spaces for collective liberation, and engage in **healing justice** and **reproductive justice** work.

Melissa is also a gifted writer, facilitator, and educator. They have spent the last couple of decades in various environmental-social justice and learning spaces on occupied Ohlone land, presently known as Oakland. They taught at elementary schools for eight years before they were inspired to author the illustrated children's book *I Am Sausal Creek/Soy El Arroyo Sausal*, a bilingual book about the rich history of Oakland told through the voice of a local waterway. Melissa has also worked as the training manager of the Center for Story-Based Strategy, an organization that strengthens grassroots movements by cultivating radical imagination and phenomenal creative strategy. It's a treat to learn from Melissa embracing so many different modalities to hold, care for, and gently guide folks through and beyond this moment of time.

Content Notes: Settler colonization, imperialism

On her family's history shaping her worldview

Three of my grandparents immigrated to New York from Puerto Rico, or Borinquen, and my parents grew up as immigrants, poor, in Harlem. Growing up, I had access to privilege because my parents were able to have a relative amount of success in the United States. I spent a lot of time grappling with how I ended up a person of Puerto Rican descent, growing up in the United States, completely isolated from my people. In order to understand who I was, where I came from, and how we came to be in this situation, I had to begin to understand colonization and patriarchy and how those things have defined my family's story.

These experiences really shaped my sense of understanding of the world and informed the work that I do today in justice-oriented spaces, whether environmental-social justice or healing justice work. Because of my fundamental belief in the goodness of people and in our right to live healthy, dignified, and joyful lives, and seeing so many people denied that, I thought, *Okay, what am I going to do about it?*

On storytelling as individual and collective healing

At the Center for Story-Based Strategy,[1] we're really homing in on how we can create spaces where people can dig deep into imagining and visioning the world that we want and deserve to live in. At times, that might look like expanding on or opening up the ways that we have traditionally thought about organizing or collectively creating new worlds. It feels inspiring to me to work with some of the tools of story-based strategy and narrative strategy to help open up some of that space for people.

For me, there's a lot of healing that takes place in storytelling. Speaking our stories to each other, being witnessed, and seeing ourselves reflected in other people's experiences is powerful. Conversely, it's really powerful to see ourselves reflected in other people's stories. I can think of many moments in my life where I felt isolated by an experience or I thought that something was wrong with me, but by giving voice to my story and connecting with other people, I realized that it wasn't just me. Instead, I could see my story as rooted in a violent system. Pulling together these threads and understanding the context helps us collectively find a way forward.

On another level, storytelling is about reclamation and right relationship. The art and practice of storytelling is a very Indigenous practice. Indigenous folks and land-based folks all around the world have really rich traditions of storytelling; that's how culture and values have been passed on. I've heard folks talk about how, in the stories their grandmothers would tell, there are teachings in there about how we are meant to live in this world. For me, the idea of storytelling is also about reclaiming those teachings that contain a lot of wisdom about how we live in balance on this planet.

Narrative strategy, meanwhile, is very much about understanding power. It's about understanding underlying assumptions and the values embedded in the narratives that we are saturated with. Because stories are one of the pieces that uphold the system. Like, if we look at the story of Thanksgiving, there are whole sets of values embedded there around whose lives are more valuable than another's. When we begin to understand the stories that we're saturated with in mainstream or dominant culture, we can intervene and change those stories, and in combination with organizing and taking action together, that's how we can actually shift power and shift the values that uphold the systems that are hurting most people and this planet.

After spending eight years teaching in elementary schools in Oakland, I decided to write a children's book. There are so many children's books that, if you really look at them, are problematic. And these are the books that we read to our kids every single day. These are stories that teach our babies about the world we live in, what's important, how to relate to each other and the world around them, and about their place in the world. The values and visions for the world that are embedded in those stories really impact people in their day-to-day life. So for me, children's books are a powerful place of intervention in storytelling.

On healing as a necessity and a right

I study and practice curanderismo, which is a holistic care system and body of traditional healing practices rooted in what we now know as the Americas and Caribbean. It really draws in a lot of different healing modalities, and my practice in particular is rooted in birth work, herbalism, and bodywork.

My healing justice work started and continues with my own healing and an understanding that, if we're actually going to create transformative shifts in this world, we have to tend to those wounds that keep us locked into old ways of being that aren't healthy or serving us or our communities. Individually and collectively, so many of our issues stem from trauma. A lot of us are holding and carrying generations of trauma on top of the trauma of living in a very violent world. If we don't work through and take care of the harm that we've experienced or have inflicted on each other or this planet, we're just going to continue perpetuating it.

I'm really passionate about this, in part because healing practices have been important for me to be able to show up in the world in the ways that I want to, but also because I believe it's deeply connected to being able to create a more just world, filled with love and compassion—a world that actually celebrates life! Having the opportunity to heal ourselves and work with each other toward collective healing is how we are going to learn to be in right relationship with this planet and to shift out of a system that is based on the genocide of Indigenous peoples, the stealing of land, the enslavement of people. We are only going to be able to create transformative change if we can get in and do the work of tending to the hundreds of years of violence that most of us are still swimming in.

For me, healing is also about remembering relationship to the earth and to the plants that nourish us so much physically but also energetically and spiritually. The plants have been here long before we were here. They have so much to teach us about how to be in this world and how to honor the sacredness of life in loving and reciprocal relationship. They teach us what it means to accept and welcome in the abundance of this earth and also take care of it. They teach us about how we need each other and how that's beautiful.

RESILIENCE TOOL SPOTLIGHT

Story-Based Strategy

Story-based strategy is a framework put forth by the Center for Story-Based Strategy to prioritize imagination, narrative analysis, and thoughtful storytelling in movement strategy. The organization has developed a set of tools to help organizers and activists unpack the stories of the opposition, identify power dynamics at play, understand audiences more deeply, and craft accessible, memorable, and evocative messaging.

It should be no surprise that the success of organizer and activist campaigns and movements often hinges upon their ability to tell really good stories that challenge dominant narratives and shape popular culture—after all, storytelling is an ancestral practice across all cultures that has been used to inform, educate, entertain, connect, and make meaning for millennia. Yet the importance of storytelling is often overlooked.

The climate movement specifically has historically suffered from a lack of resonant, imaginative, and clear storytelling. While there are undoubtedly numerous climate communicators working hard to deliver climate stories with nuance, creativity, and care, a lot of the messaging developed by environmentalists and reinforced by the media over the past several decades has fallen into the same handful of traps: it has framed the climate crisis as a faraway problem, both in time and space; it has fixated on abstract trends and quantitative data; and it has failed to connect climate change to social justice issues and everyday quality of life. The most-discussed solutions have been framed as too small, like minimizing one's own carbon footprint, or too out of reach, like passing federal legislation or inventing groundbreaking technology. The tone of climate discourse has frequently oscillated between cold, fact-focused language, extreme panic and doomism, and technocentric optimism. Altogether, dominant narratives around climate change have largely failed to mobilize the masses into action. They have instead left many folks feeling overwhelmed, mired in guilt, unsure how to act, and afraid in the face of the future.

This moment calls for a new set of climate stories and amplification of the folks who are already doing climate storytelling differently. To successfully navigate our way through and beyond the climate crisis, we will need to prioritize radical imagination and collectively envision a different way of being in the world and relating to one another and the more-than-human world. Rob Hopkins, the founder of the UK-based Transition Towns movement and author of *From What Is to What If*, writes, "We need to become better storytellers in such a way that we can, through a variety of media, give people a visceral sense of what a positive future would sound, taste, feel, and look like . . . Every vital and much needed conversation with people about the gravity and scale of the challenges we and the natural world face should also include a taste of how it could be, a story of the future we could create."[2] It's vitally important that people can relate to and visualize the future that they're fighting for.

To incorporate story-based strategy in your organizing, activism, and conversations:

- **Reflection and journal prompts:** (1) Spend a few minutes (or an afternoon) meditating on the future that you're fighting for, beyond survival or a certain level of greenhouse gases in the atmosphere. What does this just and abundant future look, smell, sound, taste, feel like? What is present, and what is not? Feel free to doodle or jot down thoughts when inspiration strikes, and keep this journal entry or piece of art somewhere visible so you can revisit it during moments of frustration, doubt, and exhaustion. (2) Take a moment to consider the stories that frequently pop up in climate change discourse. Which stories resonate, and why? Which stories feel harmful or counterproductive, and why? What are their underlying assumptions? Where did they come from? Whose stories are ignored or erased to create this narrative? How might you transform, shift, or replace those narratives? (3) There are countless media through which to convey stories, like songs, memes, slogans, speeches, films, photography, movement choreography, printmaking, painting, drawing, and more. Which media do you most enjoy working with? What could story-based climate strategy via this medium look like?
- **If you're interested in expanding your story-based strategy toolset:** The Center for Story-Based Strategy (CSS) has oodles of awesome templates to get started on your own or in community. Their ninety-minute

self-paced virtual workshop is a great place to begin. You may also consider inviting a CSS trainer to partner with your group for a workshop or long-term story-based strategy development.

- **If you (or a fellow group member) enjoy facilitating:** Lead your group (and then the larger community) in a visioning exercise. The Transition Network has a great facilitation guide to their What If visioning exercise for group and community settings (available as a PDF at www.transitionnetwork.org). The format, which asks participants to take stock of what is, create a vision for what could be, and consider how to get there, is a beautiful way to stoke collective imaginations, unite folks in common purpose, build relationships, and kick off a new campaign, initiative, or movement.
- **If you have access to financial resources:** Fund the type of storytelling that you'd like to see more of. Whether you subscribe to a content creator's Patreon platform, pay for a climate justice advocate's newsletter, donate to a favorite environmental reporting platform, or contribute to a fundraising effort to bring a storytelling project to life, there are infinite ways to support the folks who are working hard to tell new climate stories.

PATTY BERNE

SHE/THEY

"We need to make the revolution absolutely accessible"

The work that Patty Berne has done to center disability—and particularly disabled queer and trans people of color—in conversations around climate justice and resilience cannot be overstated, and this book would have felt wholly incomplete without their wisdom. Patty is widely recognized for their work establishing disability justice, which is a practice, framework, and movement predicated upon the understanding that ableism is inextricably entwined with white supremacy, colonial conquest, and capitalist domination, and therefore cannot effectively be addressed within a single-issue civil rights framework. In 2005 Patty Berne and Mia Mingus, both queer and disabled women of color, developed the disability justice framework to offer an expansive alternative to the disability rights movement, to address ableism alongside cis-hetero patriarchy, white supremacy, colonialism, and capitalism, to center the lived experience of disabled queer and trans people of color, and to fight for liberation for all.[1]

Today Patty is the executive and artistic director of Sins Invalid, a disability justice–based community organization and performance project founded by Patty and Leroy F. Moore Jr. in 2005.[2] Sins Invalid is focused both on growing and nurturing the disability justice movement and producing performances that explore themes of sexuality, embodiment, and the disabled body, challenge ableist paradigms, and offer up all-embracing visions of beauty and wholeness. Most recently Patty directed We Love Like Barnacles: Crip Lives in Climate Chaos, a multimedia performance centering the experiences and wisdom of disabled queer and trans people of color in the age of climate chaos. Patty's work is informed by decades of organizing and advocating for justice and liberation for the most marginalized, a background in clinical psychology focused on trauma and healing for survivors of interpersonal and state violence, and their own lived experience as a Japanese Haitian queer disabled woman.

Content Notes: Structural and interpersonal ableism, structural racism, death, mass incarceration

On developing a disability justice politic

I came out as queer when I was like sixteen, and I kind of came out again as a person with a disability after watching and experiencing Leroy Moore's work and becoming friends with him, and also through the written work of Essex Hemphill,[3] a Black man who had HIV. It's one thing to be functionally disabled—I've been disabled all my life. But to come out politically as disabled was a challenge because the disability rights movement was so white. It was when I met Leroy, who had a critique of the disability rights movement and community, that I was able to embrace disability as a political identity that I held. And I'm ever grateful for that.

The disability rights movement has had really important achievements, particularly in creating a political identity around disability and moving us out of thinking of disabled people as individual tragedies. However, in some ways the movement abandoned a grassroots strategy and went for more of a more formal, bureaucratic approach. The average person with a disability has more rights now, but often people with disabilities still think of ourselves as unworthy and unlovable because there wasn't really a heavy emphasis on cultural work and shifting how we understand ourselves. Yes, we need rights, but my kin also need to know that we offer a lot of brilliance, a lot of power analysis, sexiness, and love. I think that liberation really starts from seeing that in ourselves. Liberation isn't just external.

I also think that liberation must be collective, that there's no liberation of our hearts and souls and our **bodyminds** and our communities without thinking of how we're all connected to each other. That's kind of encapsulated by the idea that if anyone is still oppressed, we're all oppressed, or that liberation for one is liberation for all. I can't stress enough how all our struggles are intertwined.

All those beliefs led to the establishment of disability justice as a framework. Mia Mingus (a writer, educator, and community organizer working toward disability justice and transformative justice) and I were literally sitting in my office and debating whether or not we should call it the second wave of disability rights and how we wanted to frame this understanding of justice. And we were like, *Well, should we just call it what it is—justice?* Mia had done work in reproductive rights and had moved toward a reproductive justice framework. And I had done work in multiple justice movements. And we were like, *Why compromise?* Let's really reach for what we want, which is disability justice.

On disability justice and collective care as climate justice

I have to say very bluntly that disabled people, and particularly disabled people of color, are an afterthought in policy. Other people thinking about what we might need is pretty rare. And so when we think about how all the ways that we're marginalized are exacerbated by climate change, disability becomes an exponential multiplier. For example, we're only allowed to have so much in our bank accounts before Social Security tells us that we can't use public benefits, assuming that we are citizens.[4] How is someone supposed to evacuate on a thousand dollars or less? Especially considering the costs of travel, the costs of attendance support, or having a backup vocal communication device, or the cost of hotel rooms that are accessible to wheelchair users or folks that are hard of hearing or Deaf. For myself, as a chair user, I would likely have to get a king suite to have a roll-in shower at a hotel. And nobody has the funds for backup equipment. The medical industrial complex really takes advantage of a captive audience. Everything is bumped up in cost.

Sadly, when we look at climate emergencies like the Camp Fire in Northern California, which basically eradicated the entire city of Paradise, they said that eighty-five people died, and I immediately knew that they were disabled people or elders who didn't have funds to escape.[5] In those situations, we die. That's what motivates so much of the work of Sins Invalid and of my own work. It's my love for us as disabled people of color and knowing how valuable my community is, how rich and beautiful and brilliant we are, and wanting to make sure that we survive. I want to make sure that there's a future with us in it.

This concept of *we keep us safe* is not just an idea. There is no public safety net that includes disabled people, particularly disabled people of color and particularly those of us further marginalized by cis-heteronormative patriarchy. We really do have to lean on our own communities and our personal networks. If other people don't value us, that doesn't mean we can't look at our friend and be like, *You're amazing and valuable, and I'm gonna make sure you don't die.* We may not have institutional power, but we have the power of numbers and the power of our connections and our valuing of each other, which is basically what love looks like: valuing and respecting the dignity of another person.

On building more accessible movements

We have to balance the tension between being slow and intentional to create spaces that are inclusive[6] of all of our bodyminds, and also really acting on the urgency at hand, at the precipice of existence for our species and for many kin species. I haven't yet seen a lot of mobilizations that center or include disabled people of color, and I would love to challenge my comrades to incorporate more disability justice principles and communities of crips[7] in broader popular struggles.

There's so much more that could be done. To make mobilizations and rallies more accessible, organizers could designate quiet spaces for neurodivergent folks and have ASL interpreters for Deaf comrades. In-person mobilizations are important, and I'd also love to see other ways of participating that aren't in-person, like maybe folks who have mobility impairments could all wear a particular color and take a photo of themselves to send to a central location. I often think about the fact that more than one-third of the people who are incarcerated are disabled.[8] How could they get involved in our mobilizations? There are so many ways of including people who are in different circumstances from the typical able-bodied, young to middle-aged activist, but we have to use our creativity.

We also need to have disabled people of color not just on the ground but in positions of leadership. That's how you truly include people, not just at one level, but at every level, from local leadership to national leadership to international leadership. That's how we can crip[9] the resistance to the **anthropocene**.

On the cultural strategy imperative

I can't stress enough that capitalism is not as powerful as it is because people like to be exploited or isolated. The exploitative economy has the power of violence, but it also has the power of our culture. They've really made capitalism seem appealing through culture and through all forms of media. Corporations are good at that. They spend millions of dollars trying to capture our attention in the hopes of capturing our hearts.

So it's our responsibility to make the revolution irresistible, as my bestie Micah Bazant says, and as Toni Cade Bambara said.[10] That's the truth. We need to make the revolution absolutely accessible for people, and we need to have our vision of our wholeness and our beauty be reflective of our audiences. We need to have the truth of our fantastic bodyminds be celebrated so that we see ourselves as deserving of all that we have to offer each other, which is all the love and care in the world. And I think we can absolutely do that.

SELIN NURGUN

THEY/SHE

"Tapping into our awe and joy and inspiration is mandatory"

Selin Nurgun is a somatic coach, climate grief worker, educator, and facilitator rooted in anti-oppression and liberatory frameworks. She supports people in partnering with their bodies to show up in the world with compassion, power, and in alignment with what they care about most. As a member of the Climate Psychology Alliance, she is working with others to integrate a social justice–oriented framework into the emerging field of climate psychology. I first came across Selin's brilliant work through Instagram, where they challenged me to think in entirely new ways about how we make sense of and respond to climate-related emotions.

As a queer, femme, first-generation Turkish American who grew up straddling two cultures, Selin has an approach to life and climate action deeply informed by her heritage, as well as her background in psychology and environmental behavior, education, and communication. In their work and daily life, they center ancestral remembrance, collectivism, pleasure and joy, and deep connection with the nearby rivers and woods.

On recognizing the emotional weight of climate work

About ten years ago I was really active in the climate movement on my college campus and in some pretty big climate protests. Movement culture in the West in particular is really urgent. It's very intense. And there's a lot of gatekeeping. So there are many walls that you hit. For me, and for many people, I eventually got pretty exhausted and also discouraged from not seeing change. I started to wonder, *What am I doing? What is the point of all this?* It took me many years to realize that I was internalizing the resistance and struggle and disenchantment. I was hurting.

And so I stopped. I didn't want to be a part of anything anymore. I even stopped reading the news or keeping a pulse on the climate movement. And frankly, for folks who experience something similar, I would encourage taking that break. That's probably your nervous system saying that you've been on alert for so long and there's no way to just keep pushing through. It's okay to rest and trust that other community members have your back. We can tag-team this.

But then after a year or two, when I was really honest with myself, I realized that I belong in that space. I was like, *Wait a second, I'm passionate about this stuff. Why am I isolating myself instead of finding community around this experience?* It was a good realization. It brought me into **deep ecology** work and Joanna Macy's work. I participated in a Work That Reconnects circle.[1] It opened my eyes and made me realize that there *is* an emotional aspect to climate work. I realized that I wasn't the only person who was feeling the weight of the climate crisis and whose mental health and stability was impacted. Somatics came into my life when I was trying to find a way to bring people into the space of getting unstuck and moving through these difficult feelings. Bodywork really resonated with me. It made so much sense as the place to begin.

On why the antidote to climate anxiety isn't necessarily action

A common refrain in the climate movement is that the best antidote to anxiety is action. When I first started hearing that kind of messaging, I wasn't hugely bothered by it, even though I disagreed with the sentiment. It's just not true that jumping into action is going to absolve all uncomfortable feelings. But the more that I saw this message proliferating, I started to realize it was reflective of something bigger happening. When I hear people talking about jumping into action to combat anxiety, it sounds like an attempt to numb, to disconnect. I don't think the message is intended that way, but the subtext is that you should try to escape what you're feeling. And that makes me really nervous. Especially because the climate actions that folks are talking about are often pretty big, like joining an organization or gathering for a protest on the front lines, where there's more to do and more to take on and people often become even more overwhelmed. When people dive into action without an awareness of their body and their triggers and trauma responses, activism can easily turn into a cycle of burnout and shame. I've been there.

I want to affirm that the climate crisis is something that is happening to our nervous systems, and also that that's not necessarily a bad thing. Our bodies very wisely evolved to react to threats of safety. It's a good thing that our nervous systems are looking out for ourselves and our communities. I like to ask people to consider how we can befriend anxiety, in a way, to listen to what it's telling us. Climate anxiety is a reasonable reaction to traumatic global events and structural oppression. It's a healthy indicator that you're awake. What does it want to tell you? What would happen if you slowed down and listened to it?

The reality is that we're going to keep experiencing new versions of this world. Change is going to be constant. So why aren't we actually learning how to safely grow our capacity for discomfort and uncertainty and take steps that care for our bodies *alongside* the work? Instead of trying to get rid of our anxiety and fear and dread, I want us to focus on learning strategies to better cope with it. Coping should be a way to stay, not escape.

On centering pleasure in revolution

One of the practices that I return to most regularly is an exercise of connecting to our resilience. Start by imagining somewhere that just feels juicy and good in your body. Maybe it's a good memory or maybe it's someplace futuristic. Your practice could even be physical, like putting your feet on the ground and in the grass. Sit there and start to take note of how your body is experiencing this moment. What sensations are coming up? The more that you cultivate this practice, you start to deeply understand that pleasure and joy are always available to you. You can access those feelings whenever you want, and nobody can take them away from you.

It's critical that we take space to regularly remember and embody our big *why* behind this work. I don't do this work because I'm angry, although those feelings are very real, too. It's a desire for beauty, connection, and sovereignty for all peoples that drives me. Ultimately, the revolution is for more joy, more pleasure, for the love of our brothers and sisters and community. When we forget this and feel disconnected from our *why*, and perhaps just focus on what we are resisting and what's getting in our way, the work starts feeling more burdensome than enjoyable. And so tapping into our awe and joy and inspiration is mandatory. As Emma Goldman said, "If I can't dance, I don't want to be part of your revolution."[2]

People Power

To evade the worst possible climate scenarios and effectively address the paradigms of white supremacy, settler colonization, growth-at-all-costs capitalism, and cis-hetero patriarchy at the root of the climate crisis, power must be returned to the people. These essays are from organizers, advisors, and strategists who are growing movements, preparing folks to govern, shifting policy and resources, dismantling systems, and building new ones. They consistently demonstrate that together we can, and must, win.

SEKITA GRANT

SHE/HER/HERS

"Frontline communities have a lot of creativity and innovation that should be honored and respected"

Sekita Grant is an environmental policy and social justice strategist, philanthropic advisor, and climate justice advocate. After working as an attorney and going back to school for her master of laws in environmental land use law, she conducted legal research with the Nature Conservancy, shaped energy policy with the California Energy Commission, advised large corporations on sustainable business practices with Business for Social Responsibility, advocated for justice-centered climate policy with the Greenlining Institute, and more. I love that Sekita has gained firsthand experience in virtually every sector of the climate field. It's given her a tremendously nuanced understanding of the climate space and why it's so important to nurture grassroots movements.

Today Sekita brings her expertise to frontline communities across Turtle Island and beyond, primarily working with coastal and island communities that are historically marginalized and underresourced. She recently partnered with legendary environmental justice activist Catherine Coleman Flowers to champion policy that addresses air, water, and soil contamination in rural Southeastern communities. She has worked with the Institute for Climate and Peace in Hawaii in developing a new climate resilience initiative and has worked with La Maraña, a participatory design and planning nonprofit in Puerto Rico, to pivot into just recovery work after Hurricane Maria. As the vice president of programs at the Solutions Project, an organization that moves resources to grassroots innovators and frontline leaders, Sekita oversees grant-making, policy research, and impact strategy support.

Content Notes: Structural racism and classism, settler colonization

On navigating barriers and the need for intersectionality in environmental spaces

I am a nature nerd. I always have been. I love it dearly. When I was younger, I would obsess over how we could clean the air. When I was eight, I even dreamt up this idea for a giant vacuum cleaner for the sky.

As I got older, I went to law school and started working as an attorney. At one point I was working at a bigger law firm and wasn't really enjoying the work, so I thought maybe I'll do some volunteer work for some groups that I had come across doing beach cleanups. I was volunteering with the Surfrider Foundation (a nonprofit environmental organization focused on water quality, equitable beach access and preservation, and marine and coastal ecosystem health) a lot, and since they're active legally and politically, I asked my law firm if I could do some pro bono work for them. The firm said they would look into it. When they got back to me, they said that I couldn't work with them because it would be a conflict of interest with our clients who were doing real estate development. At that moment, I was like, *Okay, I need to make a change.* So I went back to school to study environmental land use law.

And then when I started working in environmental spaces, I was like, *Wait, nobody looks like me. What is this?* Anytime that I started to look or push for the strategies that were speaking to the needs of people of color and low-income people, it was hard to find. I didn't see a lot of access points and engagement. That's not to say that it wasn't there at that time; that's just where my journey was. There have been folks doing this work for decades that I just wasn't plugged into, but I was also working in areas of privilege and power. That's when it really clicked that the work that I was really interested in was at the intersection between environmental and social justice work. So I started to sink in and open my eyes to that world. And once you see all of the very blatant racism and economic injustices occurring within climate impacts, but also in climate solutions, you can't unsee it.

For instance, I'm on a listserv with a bunch of electric vehicle (EV) folks, and somebody recently sent out the exciting news, I guess, that there is now an electric version of the Hummer vehicle. I had to sit with that for a second. I get it. I know people want to make that technology accessible and capture a bigger market share. I worked a lot on EVs, and I see that they have the potential to be a very strong solution if deployed in the right way. But when I worked in that field, I saw a lot of money for climate change solutions being diverted to EVs, away from other solutions. We were spending money on a technology that many low-income and moderate-income people wouldn't be able to afford or access, and it diverted money in ways that ended up supporting folks who already have money.

Along the same lines, when I was in Bay County, Florida, after it got hit by Hurricane Michael in 2018,[1] I watched the disaster response play out in a really inequitable way. I went to some of the local government meetings where they were discussing the hurricane relief response, and it was a lot of white men. Sure, these meetings are open to the public, but I had spent time at the building where you get your disaster response resources, and it was a lot of moms, a lot of young kids, people living out of their cars. They do not have time to go to these meetings. They are in complete crisis mode, and they don't have adequate representation in decision-making spaces. The conversation being had with the local government was really focused on getting businesses up and running. Meanwhile, the population of young people that were unhoused *tripled* after the hurricane.

If you're one paycheck away from missing your rent, an unpredictable event can send the course of your life into some really deep hardship. If you don't speak English, or you're illiterate, or you don't have a nonprofit to help support you in filling out a FEMA form, which are really technical, it can be impossible to receive disaster assistance. There are so many people completely left out or not prioritized in the wake of disasters. In fact, there are studies that show how white communities will overall gain money after a big disaster while Black communities lose tremendous amounts of money.[2] There's a major racial wealth gap issue in the disaster response world that has yet to be fixed.

On policy cocreation as a powerful climate resilience tool

When we talk about climate policy, it's important to first name that there are a lot of existing policies that interfere with communities' ability to be resilient. When immigration policies separate families and freeways are put up through thriving Black communities and rent prices skyrocket and people are displaced, that inhibits a community's ability to be resilient.

So it's not always, *How do we go in and save people?* or *How can we help this community be more resilient?* Part of it is, *What are we doing that is causing climate change and also perhaps undermining the ability of the community to be resilient in general?*

I also want to point out that there's an important world of resilience that many climate policies, and policies in general, miss. Frontline communities have a lot of creativity and innovation that is not always seen or heard. Many marginalized communities can speak to the fact that to survive, they have had to be resilient. For more than 400 years, Indigenous communities in the Americas have had to figure out how to be resilient to survive. That shouldn't be the case, of course, but that fact should be honored and respected. And one of the ways to do that is through cocreating policy as much as possible, at every corner. Create policy with community, meaningfully.

Different parts of the United States have started co-designing policy more, or doing more participatory budgeting, which is a big part of policy. It makes sense that you have community members in the room who are actually going to be impacted by the resources. They should be co-designing the programs and have voting power. For one, it's the right thing. Second of all, it's activating, so instead of having one decision maker and five folks on a committee, you now have hundreds of people who are excited, behind the program, and are going to be engaged and attending meetings. You have collective energy moving toward ensuring that the program and the work are successful. Third, you're engaging the folks who have been most harmed by climate change and who have the most to lose. They are motivated to get this right and can bring a really nuanced understanding of how the program might succeed or fail.

This work cannot be done in a carbon silo. If you don't talk to people and seek their wisdom in ways that are just, if you're not working with communities, then I know you have the answers to the test wrong. We can't just Hummer EV our way out of this. And even though eight-year-old Sekita would have loved it, we can't just vacuum the sky.

RESILIENCE TOOL SPOTLIGHT

Participatory Budgeting

Participatory budgeting, or PB, is a simple but transformational citizen engagement process where community members decide how a portion of a public budget is spent. The concept emerged in Porto Alegre, Brazil in 1989, in a moment when the city was growing quickly and struggling to meet the basic needs of its residents. The experiment proved to be hugely successful. A decade later, infant mortality had declined by 33 percent, the number of schools had quadrupled, health budgets had tripled, and resources for basic infrastructure, like roads and sewer and water connections, flowed into the poorest neighborhoods.[3] Since then, more than 7,000 cities around the world, including 29 in the United States, have adopted a participatory budgeting process.[4] Hundreds of millions of dollars have been allocated by residents through a community decision-making process. The movement is growing, and for good reason.

If budgets are moral documents, then "participatory budgeting is a way to ensure that public budgets are living, breathing reflections of community values so that they can be functional and effective moral documents," explains Shari Davis, coexecutive director of the Participatory Budgeting Project.[5] Davis elaborates that the so-called United States is currently a representative or indirect democracy, where elected officials create and vote on policy on their constituents' behalf, but practices like PB can move the country toward a direct democracy, where people participate in government in meaningful ways beyond just voting in elections.

In the context of the climate crisis, a PB program allows residents to determine and actualize meaningful climate solutions that work best for their community. Particularly as national and international leaders continually fail to act at the speed and scale necessary to ward off extreme climate change scenarios and keep communities safe through escalating disasters, PB programs at the local and regional scale can empower communities to launch projects that will help minimize their environmental impact, prevent harm from extreme weather, and bolster recovery efforts during and after disasters. Because community members have the opportunity to vocalize their concerns and priorities on a regular basis, and not just during election season, PB programs can help communities be nimble and responsive in the face of changing conditions.

The PB process is relatively straightforward. First, a segment of a municipal, institutional, or school-district budget is allocated for PB. Communities generally earmark between 1 percent and 15 percent of an annual budget, or an amount substantial enough that community members are incentivized to participate and will be able to experience material differences from the process. (In Paris, France, for instance, the mayor allocates 5 percent of their annual budget, or 100 million euros, toward PB.) The budget is usually taken out of capital discretionary funds, or funds that aren't already committed to everyday expenses and can be used at the discretion of public officials, school administrators, and the like. Next, the community helps design an inclusive PB process. Through community meetings and online platforms, people brainstorm projects that they'd like to see funded. It's an opportunity for collective radical imagination. Then volunteers help calculate estimated project costs and flesh out ideas into full project proposals. After community members get the opportunity to compare potential projects, the list is narrowed down. Finally, community members vote on their favorite projects, and the projects with the most votes are funded and brought to life. The projects that don't get funded still serve as an important source of feedback for public officials or administrators.

To get involved with the participatory budgeting movement in your community:

- **Research prompts:** (1) Does your city have a participatory budgeting process in place? If so, what stage of the annual PB cycle are they in, and how can you get involved? (2) Take a look at past and proposed budgets for your city or municipality, and particularly the capital budget, which typically includes long-term investments in facilities and infrastructure.
- **Journal prompt:** If budgets are indeed moral documents, what does your city's budget say about your community's values, goals, and priorities? How would you reallocate your municipality's proposed budget to maximize for resilience, justice, and collective well-being?
- **If your city or district already has a participatory budgeting process in place:** That's amazing! There are myriad ways to get involved, from sharing a new project idea of your own to showing up as an enthusiastic voter. PB processes are often largely run by volunteers, so if you have a passion for facilitation or a background in accounting or developing project proposals, your skills can likely be used.
- **If your city or district does not yet have a participatory budgeting process in place:** Advocate for one. Talk to neighbors about what PB entails, reach out to your elected officials, or start a petition to bring PB to your municipality. If you hold a staff or elected position in local or regional government, consider partnering with the Participatory Budgeting Project to start a PB program from scratch. This group can help with program design and planning, training, technical assistance, and more.
- **If you're involved with a group with a budget beyond government spaces:** Consider applying PB practices to a portion of the group's budget. As Shari Davis points out, PB is a tool that can even be brought into the home. Movement groups, extracurricular clubs, nonprofits, and businesses can utilize PB to democratize decision-making, get important feedback, come up with innovative ideas, increase engagement and

retention, and demonstrate appreciation. PB is also a fantastic tool for schools, from elementary schools to universities. If you're interested in bringing PB to your school as a student leader, administrator, teacher, or parent, the Participatory Budgeting Project has a great step-by-step guide including lessons, worksheets, and more.

- **General best practices:** PB has the potential to be radically inclusive, but it's not a given. At every stage of PB design and implementation, consider how to maximize accessibility for all peoples and how folks most excluded from decision-making spaces can be centered. For instance, Boston's Youth Lead the Change program specifically empowers youth, and particularly low-income youth, youth of color, and immigrant youth, to allocate one million dollars through a PB process.

MARA VENTURA

SHE/HER/HERS

"No true progress in climate justice is going to happen moving forward if the work isn't led by care workers and domestic workers"

Mara Ventura is a savvy, heart-driven organizer committed to creating space for undocumented and essential worker communities to lead and utilize the tremendous power that they already have. She cut her teeth organizing with the Oregon Student Association and as the co-founder of her university's first Coalition Against Environmental Racism, a group that gave her ample space to explore the environmental injustices perpetrated by oil companies across the so-called Americas. These experiences helped Mara deepen her understanding of her lineage as a transnational, transracial adoptee from Bogotá, Colombia, and crystallize the connection between workers' rights, immigrant rights, and environmental justice movements—and fall in love with collective action.

In 2017, after nearly a decade in the labor movement, Mara joined the Jobs with Justice team based in the North Bay, a region north of San Francisco that is home to nearly 1,000 wineries, most of which are powered by farmworkers born outside the borders of the so-called United States. A few months into the job, the Tubbs Fire tore through Napa and Sonoma Counties, killing twenty-two people, incinerating dozens of city blocks, and displacing thousands of people. Mara, who was among those evacuated from their homes, knew that the area's many undocumented residents would have the least access to resources and services to recover in the wake of the fire, so she leapt into gear from the shelter where they had evacuated to. In partnership with other justice-oriented community organizations, UndocuFund for Disaster Relief in Sonoma County was born to ensure that undocumented families impacted by the fire would have access to unrestricted recovery funds. While their fundraising efforts were certainly commendable, what set Mara and fellow partners apart in their approach to UndocuFund was how they redistributed resources and strategized with community members to create meaningful, long-term changes across the region. Their story is a profound example of how to respond with deep intentionality in the aftermath of disaster.

Content Notes: Environmental injustice, imperialism, structural classism, and racism

On the creation of UndocuFund and the Alliance for Just Recovery

In October 2017 the historic Tubbs Fire tore through my community. My partner and I, along with thousands of others, were woken up at two in the morning and told to evacuate. Three days later, the Graton Day Labor Center, North Bay Organizing Project, and I, representing North Bay Jobs with Justice, launched UndocuFund.

It was a wonderful partnership for a few reasons. For one, Sonoma County, like most of the country, is way more flooded with services than it is with organizing for marginalized communities. I appreciated being around a table with other organizations that were deeply rooted in providing opportunities for leadership development and grassroots, direct-action organizing. It allowed us to be fully aligned with the intention to get people immediate funds with as few barriers as possible, but also to build pathways for folks to create change around the systemic barriers facing our communities.

So during the intake process, we made sure to capture data around whether people were renters, which neighborhoods they lived in, which industries they worked in, how much they were making per hour, etc. We also stayed engaged with folks through listening circles, text messaging services, surveys, and focus groups, and we learned more about the barriers they faced down the line, how far their checks went, and what else they wanted to see. This information helped North Bay Organizing Project to create a tenants' union, hold Know Your Rights trainings, and organize folks around anti-eviction ordinances. It enabled the Graton Day Labor Center to grow their domestic workers group and raise minimum pay standards for anyone hired at the center. And it helped us at North Bay Jobs with Justice to lay the groundwork to raise the minimum wage across four cities. All of this came out of our ability to think about UndocuFund as more than just a service and a check, but also as a medium to change the conditions that created the need for UndocuFund in the first place.

While we were really intentional about gathering the data necessary to help inform organizing strategy, we were also very conscious about creating an intake process that was highly sensitive, trauma-informed, and accessible. One of the best decisions that we made was partnering with community clinic healthcare workers, like nurse practitioners, doctors, and medical assistants, who already primarily worked with undocumented folks. They helped train the volunteers and shape the intake process in a way that wouldn't replicate people's trauma. They made sure that the first thing that we said when collecting people's information was, "Just by being in this room, you deserve money, you will get money, and you are worthy. There's nothing you need to prove today. I believe you. My job is just to capture a little bit of information that will allow us to get you a check as quickly as possible and understand where you've been impacted in the community so that we can help create some change." Starting with that message right away ensured that people felt like they didn't need to prove anything. They didn't owe us their trauma.

We also wanted to get ahead of this situation where undocumented folks don't want handouts. Racist messaging from right-wing conservatives has deeply impacted people from receiving services that that they are fully qualified for and deserve. We even heard from healthcare workers that many undocumented women decline the free mammograms that they are provided because they aren't comfortable with the idea of a handout. So instead of asking folks how much money they would need to get through the next month or two, which we knew that many folks would minimize, we went through the basics of their finances: what was their average phone bill like, how much of the food in their refrigerator spoiled during the power outage, how much did they typically spend on food each month, etc. That way, we could get a real sense of the amount of money that we would need to move to adequately meet their needs.

Three weeks after we started UndocuFund, one of the cofounders of North Bay Jobs for Justice and I also started the Alliance for Just Recovery as a means to address labor issues and environmental justice all at the same time. We knew that the Tubbs Fire was an open invitation for disaster capitalism and **low-road contractors** to come into Sonoma County, and we felt an intense urgency to start organizing people immediately. We were able to mobilize actions from our local supervisors all the way up to the governor of California to kick low-road contractors out of Sonoma County and ensure that union workers were able to do the important work and that there were also pathways for folks who were undocumented or immigrant workers to access these jobs safely and fairly.

On the roots of environmental destruction in her communities

The Tubbs Fire felt like such a big deal at the time. Now, four years later, we've seen many more record-breaking, historic fires and natural disasters. We are clearly seeing the impacts of decades of environmental decimation and destruction that are the direct products of decisions that elected officials made on behalf of major corporations who still have immense power across the world.

I think constantly about how terrible it is that Western corporations have come into our communities, particularly in Latin America, and funded terrible dictatorships, created major environmental disasters, and pushed our communities to flee to places like the United States, and now so many of those folks are going through the same thing here, experiencing the same environmental disasters. What we could offer through UndocuFund feels like nothing considering that so many folks have experienced this dual trauma, both at home and in this country, and then are told by the federal administration that they don't qualify for FEMA funds or mobile homes.

Moving forward, it's only going to get worse. The kinds of natural disasters that our communities will have to face are going to wear down people's abilities to survive and the resources that we have. As proud as I am of some of the work that we've done, I also feel frustrated that the burden continues to fall on the people to try to support one another, when the responsibility should fall on the perpetrators of the climate crisis. Ultimately our government is responsible, because corporations will do what the government allows them to do, and the government has rolled out the red carpet for them for years—even now, and even as mainstream politics moves away from this idea of climate change as a hoax.

On the power of domestic workers and undocumented workers

No true progress in climate justice is going to happen moving forward if the work isn't led by care workers and domestic workers, who are primarily Black and Brown women, because they are oftentimes the most impacted by systemic racism, sexism, environmental destruction, and the attack on immigrants, and they have shown, time and time again, decade after decade, that they have the expertise and ability to make systemic change. It's no coincidence that when they are given resources, you see real change.

When I hear people talking about the victimization of undocumented workers or their inability to organize, I remind them that there are undocumented workers, and particularly women, who walk out on strike, with no rights or contract or union, who win every day. They go into legislators' offices, they get arrested, they practice civil disobedience, and they win. They may not win everything. It's a movement, it's a process. But they have been organizing and winning. They have tremendous power and know how to wield it.

On the complicated nature of resilience

There are some beautiful intentions around honoring the resilience of communities, because we want to acknowledge that people are still showing up and doing what needs to be done, given all the systemic classism and racism and poverty and other things that they've had to endure. At the same time, I want to see my community not have to be resilient. I hope that our communities don't constantly have to be bouncing back from something. I want to know that the work that we're doing today ensures that there are seven more generations that get to live and enjoy the world and that their lives don't have to be marked by the ability to live through the trauma. Because there are also so many people who are not bouncing back from the trauma, who are being killed by the police, who are taking their own lives, who don't have support to address their mental health issues, who aren't being protected. I hope the work that we're doing means less of that.

RESILIENCE TOOL SPOTLIGHT

Disaster Collectivism

Disaster collectivism, a phenomenon coined by author and historian Rebecca Solnit in her 2009 book *A Paradise Built in Hell*, describes "the sense of immersion in the moment and solidarity with others caused by the rupture in everyday life" during and following a crisis.[1] After analyzing decades of major calamities, like the 1906 earthquake in San Francisco and a 1917 cargo ship explosion in Halifax, Solnit found that, time and time again, neighbors turned to one another to meet collective needs and build new structures of care and support. In a deeply optimistic portrayal of humanity, she suggests that people are at their most altruistic and caring in the wake of crisis, and that while disasters are tragic, the collapse of normality can allow for new forms of civic engagement and public life to take root.

Disaster collectivism is perhaps the most effective antidote to disaster capitalism, a sinister phenomenon labeled by author Naomi Klein in her 2007 book *The Shock Doctrine: The Rise of Disaster Capitalism*, which refers to the way that private entities exploit political, economic, and ecological crises to strengthen their own interests, consolidate control, and erode public infrastructure. Klein illuminates a pernicious pattern: after a shock occurs, communities are disoriented and distracted, democratic norms are suspended in the name of addressing the crisis at hand, and politicians ram through a corporate wish list as quickly as possible under the guise of recovery. Samantha Montano adds in her 2021 book *Disasterology* that some aid will typically be delivered along racial and class lines, while those most impacted—BIPOC and low-income communities—are generally left dependent on nonprofits and the fickle attention of national media to receive help. She explains, "While response reveals inequality, recovery reinforces and even deepens it."[2] So many folks who survive the initial trauma of the disaster must then contend with endless bureaucratic hurdles to obtain relief and assistance, along with gentrification and displacement, poor living conditions, lack of access to food, water, and other human rights, sexual violence, domestic abuse, and more.

However, when people respond to crisis by sharing resources, cultivating lasting connections, and organizing, it's possible to both resist disaster capitalism policies and programs and propel people-powered movements forward.

The post-disaster camaraderie and solidarity that Rebecca Solnit writes about is not a given, though. During the Covid-19 pandemic, for instance, mutual aid networks proliferated, but there was also widespread resistance toward acts of collective care that would help keep fellow community members safe, like mask wearing, social distancing, and getting vaccinated. Communities must consciously commit to the collective good amid crisis, or else it's very possible that the quintessential American value of individualism will win out. Communities are much likelier to do so if they have been exercising collectivism muscles and cultivating relationships long before disaster strikes.

To help nurture disaster collectivism in your community:

- **Journal prompts:** Make a list of communities or networks you're a part of that you could help mobilize, communicate with, and/or receive support from during or after a crisis.
- **If you don't already know your neighbors:** Get to know them. Depending on where you live and what makes sense for you, this could look like organizing a potluck or block party or simply knocking on neighbors' doors and introducing yourself (perhaps with baked goods or garden goodies in tow). Once you've developed some rapport and trust with neighbors, you can strategize more specifically about how you can start preparing for possible disasters together. A wonderful place to start is

by making and distributing a list of names, contact information, specific needs during potential emergencies (e.g., neighbor #1 can't drive and will need a ride, neighbor #2 uses a respirator and will need access to a backup power supply if the grid is shut off, etc.), and community assets that might support your community during a crisis (e.g., backup generators, spare fridges, freezers, or root cellars, food and herb gardens, etc.).

- **If you love weaving and nourishing community:** Consider how you (and other pals) might bring folks together on a regular basis around a shared identity (e.g., QTBIPOC folks or new parents) or interest (e.g., a hobby, sports team, book genre, or area of activism) to experience camaraderie, joy, physical and spiritual nourishment, and sense of safety. There are countless examples of folks cultivating powerful communities through regular soup nights, dance parties, open mics, and game nights, whether at a local bookstore, bar, café, neighborhood park, or a friend's spacious living room. The key is to bring together folks who might not already know one another, so attendees have the opportunity to expand their social networks. In moments of crisis, you can then utilize your event listservs, social media platforms, and group texts to check in on one another and coordinate support in preparedness, relief, and recovery efforts.
- **If you have access to a platform:** Use it to demand an equitable, just, and effective emergency management response and to amplify stories and action items during and after a disaster (and especially the disasters that aren't garnering as much mainstream attention). Samantha Montano writes in *Disasterology*, "The media is one of the most important participants in emergency management. Survivors depend on it for information about the disaster itself—if they need to evacuate, how to stay safe, where to find shelter or food . . . Those outside the disaster depend on media to tell us what is happening, if our family and friends are safe, and how we can help. Furthermore, the media is supposed to hold emergency management accountable. When help doesn't arrive and government officials fail to respond

effectively, it is the responsibility of media to illuminate their failures. Accurate, appropriate, and timely disaster coverage literally saves lives."[3] Whether you're a journalist, run a large social media account, or have a newsletter, blog, or podcast, remember that it can be an effective tool in disseminating timely information, raising funds, and holding government and nonprofit entities accountable.

KAVAANGSAAR AFCAN

SHE/HER/HERS

"It's not just an issue for polar bears—it's our livelihood"

It's imperative that young people, and particularly those most impacted by injustices, are centered in climate and environmental movements, as they will be disproportionately affected by the consequences of today's actions in the decades to come. This idea is something that Kavaangsaar Afcan innately understands. Since she was in high school, she has been a passionate champion for young Alaskan voices, the preservation of Indigenous lifeways, and the health and safety of human and more-than-human kin across the region. Kavaangsaar blossomed into a seasoned climate justice leader and community organizer under the tutelage of Alaska Youth for Environmental Action (AYEA), a program that inspires and trains youth leaders across the state to lead on environmental issues. AYEA helped her gain the skills and confidence to lobby state senators, organize community members against Pebble Mine,[1] bring local, culturally appropriate meals and recycling to her school, and facilitate co-learning among her peers.

Today, about a decade after Kavaangsaar attended her first AYEA Civics & Conservation Summit, she is leading the program. As the Alaska Center's AYEA program manager, Kavaangsaar builds the scaffolding for the same trainings, community events, and youth-led community action projects and campaigns that instilled in her the power of grassroots organizing. She also creates pathways for youth to engage with the Alaska Center's myriad clean energy, river restoration, salmon protection, and political engagement campaigns, so that their activism can smoothly transition beyond high school.

Content Notes: Climate injustice, structural racism, and classism

On the climate impacts on her hometown villages

I grew up in three Yupik villages on the Yukon Delta, right where it empties out into the ocean. Alakanuk, Emmonak, and Nunam Iqua. *Nunam Iqua* actually means "the end of the land." The biggest of those three villages is 700 people. They're very rural. There are no roads out to those areas. You have to take a boat or a plane to send supplies or travel for medical purposes or anything like that. And so it's expensive to buy fresh food.[2] When I was a kid, a bag of grapes was $15. And that was for moldy, soft grapes. Similarly, a shelf-stable gallon of milk cost $36. So eating cereal would cost a lot. If you had the means to buy these things, I guess you were lucky, but it still wouldn't be enough.

For people who aren't used to really rural communities, and especially Alaskan rural communities, the thing to know is that our food security is entirely based on the land that we live on. So too are our spiritual traditions, our livelihood, the way that we interact with the world. We are hunter-fisher-gatherer communities. Summer is where it's at for us. We go moose hunting. We fish for salmon. We go out for days on end to pick berries to store for the winter. To say we need the land would be an understatement. We're a part of it and it's a part of us. It's integral. It's economically integral, as well, which I think Westernized folks understand a lot more, anyway.

For the climate crisis to be impacting Alaska as much as it is, at the rate as it is, it's really scary. It's changing the patterns that we have historically known for hunting or fishing. It's thinning our river ice, our sea ice, our lake ice. We need the river ice to travel between villages, to get groceries, to hunt, to ice fish. I can't tell you how many people have fallen through the ice in recent years because of the thinning of the river. It's life-threatening very immediately as well as long-term.

So too with permafrost melt and erosion; the foundations on which we live are falling through. Our houses, our schools, our community buildings. It's wild to hear an elder speak about it, because they'll say, *We built all these structures so far away from the river, and now they're falling in*. So many places need to be relocated because of this. With Alakanuk especially, which is one long village placed all along the river, there's been quite a few houses that have had to be picked up and moved. One of my relatives used to live along the riverside and I would go over there to play when I was a child. Now they live behind somebody else's house. That wasn't an easy move.

So it's frustrating. Our lives are in danger. Our homes are in danger. Our food is in danger. All three villages are fisher communities, and the main thing we fish for is salmon. Because of warming air and warming winters, the salmon have a hard time returning to the rivers they always returned to. If you ever run outdoors on a hot day and it feels so much harder than usual, it's like that, except you might be running for twenty minutes or something, and the salmon have to swim for miles and miles and miles without stopping. So when and how we're able to collect food, whether we're able to do so safely, and how much food we're able to collect is all changing. And will continue to change. It's a threat to the livelihood that we have always had and still need today.

On her journey into this work

When I was in the third grade, my teacher gave us a watered-down, digestible article to read called "Climate Change and Polar Bears." At the time, polar bears were all the media cared to talk about, it seemed. But for me, living in Alaska, I was like, *That's right by my home!* Whereas most of the country saw those headlines and were like, *I guess it doesn't matter,* it immediately affected me. I had this feeling of doom. If you're familiar with the idea of existential dread, or perhaps existential fright, it felt like that every day. I was just a little kid, terrified that my community wouldn't be able to survive. I also didn't believe it, in a sense, because I thought that with something this big, surely adults would have to come together and fix it, at least by the time I grew up. Folks would eventually get their stuff together! But they didn't. And as I continued growing, I realized that somebody must do something. And when I got to high school, it seemed that it needed to be the folks from the villages that I lived in, the people closest to the problem, who weren't being heard, and whose perspectives weren't represented. Because it's not just an issue for polar bears. It's our livelihood.

In high school, I got involved with Alaska Youth for Environmental Action (AYEA). My sophomore year they flew us to Juneau for a summit and we learned about civic processes like how a bill becomes a law, who represents our communities, how to speak to representatives and write letters to the editor, the role that media can play in putting pressure on representatives to make change that works for you, etc. Before that, I felt like I had no way to make people know why these issues were important and, specifically, important to me and my family. In that space, I finally felt like, *This is the path that I can take to make sure that my community is taken care of.* So from there, I got more and more engaged with AYEA and was invited to join the statewide board. I eventually got a job at the Mobilization Center as a canvasser. I campaigned against Pebble Mine. I got a job community organizing with the Alaska Center. Along the way, it didn't seem like there were really any decisions to make. This was just what I needed to do.

Today I still work as an advocate against Pebble Mine—we still don't want it here and never did! I'm also engaged with a project called Solarize where we help folks get discounts for solar installation on their home. I facilitate learning about the Just Transition framework and lead weekly "Climate, Care, and Community" conversations where we explore different topics and make sure folks are informed with what's going on around the state. The foundation of my work is talking to people. I make phone calls and write emails and meet people over webinars and try to follow up with them afterward to keep them engaged. Before the pandemic, I did a lot more door-to-door organizing, too. I ask people what's on their mind, how they're doing, and what they're experiencing in their community. It seems super simple, but you can't skip this step. My role as a community organizer is so focused on building relationships because there would be no community without the people in it.

OLIVIA JUAREZ

THEY/SHE

"We need to go slow and recognize that a lot of the most important work is invisible"

Olivia Juarez, a lifelong resident of the Goshute and Eastern Shoshone lands of the Greater Salt Lake region, has made it their mission to protect the mountains, deserts, and rivers of Utah and make public lands more accessible for all people. A love for the outdoors and a fierce commitment to defend the more-than-human-world were instilled in Olivia from an early age. But it was when they fulfilled a college service requirement by canvassing in Rose Park, a community located near oil and gas refineries, where they spent hours learning from the residents about how the very refineries that employed their loved ones also impacted their air quality and children's health, that Olivia was cemented on a path of climate justice and wilderness protection organizing, honoring all the complexities of both movements.

When we spoke in the summer of 2020, Olivia was the Latinx organizer with Southern Utah Wilderness Alliance (SUWA), a longstanding nonprofit that was founded to defend the region's red rock wilderness from oil and gas development and other unnecessary or harmful construction. In this role, Olivia built people power to protect the 8.4 million acres of wilderness in Utah without protection, resist fossil fuel projects, and increase safe, equitable access to public lands, particularly for Latine and immigrant communities. I so appreciated the way that they were decolonizing conversations around conservation and rethinking what it means to be a good steward of (stolen) land in the twenty-first century. Today, Olivia serves as the public land program director for the national climate justice nonprofit GreenLatinos and is a Latino Outdoors outings leader. Olivia has been recognized by organizations like the National Parks Conservation Association for their exemplary work nurturing Latine joy, healing, and leadership in the outdoors.

Content Notes: Structural and institutional racism

On organizing Latine communities within the wilderness protection movement

As a Latine community organizer with Southern Utah Wilderness Alliance, I meet community members where they're at. My goals are very similar with other community organizers in the wilderness space: to raise awareness and promote engagement to protect land from extraction and unnecessary and harmful development. I approach organizing a bit differently, however, because there are a ton of Latine folks in Utah who are nature deprived, let alone who have never been to a national park or the wild Bureau of Land Management (BLM) lands of Southern Utah.[1]

When the Southern Utah Wilderness Alliance started gaining traction in the 1980s and 1990s, their work only reached white people who were recreationists. Because white communities had the privilege to access the landscapes of Southern Utah, they subsequently fell in love with and wanted to protect the land, and so SUWA was able to gain national power. Latine communities and communities of color didn't have the same access, and so now wilderness activism is dominated by a specific group of people that had the opportunity to come to know and love these lands.

A lot of Latine folks in Utah may be the sons or daughters of immigrants, or they might be immigrants themselves, so there's a process of getting them comfortable with the idea that these public lands belong to them. Even if you're not a citizen, everybody living in America can make public comments about BLM lands, go camping, and enjoy the wilderness. So I spend a lot of time trying to get people outdoors, and especially to the red rocks wilderness. Even if we're just getting folks outdoors in their neighborhood, though, the point is to show folks that there are healing opportunities in nature nearby. We try to help folks connect the dots, that when the BLM lands are mined for oil, it ends up in the refineries in our neighborhoods, which pollute the air we breathe.

It wasn't until I was twenty years old that I was able to get to Southern Utah, and I was born here. It changed my life. I went on one trip through an alternative fall break at the University of Utah, and now, several years later, I'm working for an organization to protect those exact same lands. So we need to get more people outside, and we need to focus resources toward groups of people that have been left out of this conversation.

We need to be intentional about building relationships with historically marginalized communities with awareness, focus, love, and attention. We also need to go at the speed of trust, which makes this a really long game. For decades, communities of color were ignored or an afterthought in modern American environmentalism. After so many years of not being prioritized, there's distrust. We need to go slow and recognize that a lot of the most important work, like attending community meetings and consistently showing up, is invisible.

On cultivating Latine environmental leadership with Casa Quetzalcoatl

I'm also a co-lead of the Environmental Chinampa of a Utah-based organization called Casa Quetzalcoatl. Society often talks about grooming a "pipeline" of talent for leadership positions, but *pipeline* connotes an extractive system that we don't want. Instead, when we conceptualize the nurturing of Latine leadership in Utah, we like to use the metaphor of the chinampa, a term that comes from the floating gardens that the Aztecs cultivated in México.

In this metaphor, if the Latine folks are the fruits and vegetables of the floating gardens, then we must carefully consider all of the nutrients, pollinators, and sunlight necessary to grow the fruits and vegetables, or to develop the leadership. We think in terms of the entire ecosystem. We're focused on ensuring that Latine folks in Utah who are somewhat interested in environmentalism have all the access that they need to volunteer opportunities, internships, mentorship, and other forms of learning. We want to connect them with every experience they need to build a career if they'd like to, with the ultimate goal of building Latine leadership in many different sectors in Utah.

On wilderness preservation

I think there's a misconception that wilderness preservation means setting aside the land and leaving it alone. In reality, anywhere you go in Utah, you can see how people have been a part of the ecosystems for millennia. Designated wilderness areas keep fossil fuels in the ground, but wildlands that are co-managed using traditional ecological knowledge have the capacity for greater resilience. I'm really hopeful about the future of Indigenous peoples applying traditional knowledge to manage our forests and our deserts and our mountains. In places like the Bears Ears National Monument, the intertribal Bears Ears Commission is co-managing the area with the Forest Service and the Bureau of Land Management. As the filmmaker Angelo Baca puts it, they're bringing thousands of years of observational data to the table, and they have the knowledge and skills to take care of the areas they call home.

SONA MOHNOT

SHE/HER/HERS

"Climate impacts affect people of color disproportionately and unequally—that is why our policies cannot be race neutral"

While it has proven excruciatingly difficult to pass meaningful and just climate legislation at the federal and international levels, Sona Mohnot is demonstrating the value of focusing on bold policy change at the local and state levels. She brings a master of laws in natural resource and environmental law and lived expertise to her role as the associate director of climate equity at the Greenlining Institute, a grass-tops[1] public policy organization with a mission to meaningfully transform the material conditions of communities of color in California and beyond. In this position, Sona is helping create the blueprint for good climate policy rooted in racial and economic equity. She has authored policy-shaping guidebooks, identified opportunities in legislation to advance equity, trained policymakers and agency staff at the state level, worked with local governments to implement the Institute's equitable climate action framework, and helped shape new strategy for the State of California.

Sona was also recently appointed by the governor to be a council member on California's Integrated Climate Adaptation and Resiliency Program, which was created to coordinate comprehensive and intersectional responses to climate impacts across the state. Additionally she serves on the advisory board of the California Resilience Partnership, a public-philanthropic effort to expand resilience projects across the state, and is a member of the California Climate Insurance Working Group, a group exploring the role that the insurance industry can play in addressing climate risks and mitigating the disproportionate risk that the most vulnerable communities carry.

Content Notes: Structural and institutional racism and classism

On the imperative for an intersectional climate justice movement

Climate justice is racial justice is economic justice. We have to think about these movements not as siloed issues but as very much intersectional. The reason for that is because our country was founded on racism and white-supremacist ideas, from the genocide of Indigenous peoples to slavery to Jim Crow laws, to the present-day ICE detention centers and police brutality. We have never really had a real reckoning to address those injustices and the harm that they are still doing to communities of color today.

For instance, if we consider redlining, which is one of the policies that instigated the formation of the Greenlining Institute, it is an overtly discriminatory and racist policy. Bankers, governments, and realtors worked together to proactively prevent investments from going into communities of color. They would not invest in those places because the presence of Black and Brown people was seen as a credit risk. As a result, communities of color were underinvested in for decades, and that's part of the reason why there is still such a huge racial wealth gap today. During a time when a lot of white families in America were building wealth to buy homes and fund their children's education, those opportunities were effectively not available to Black and Brown families. That general wealth does not exist in a lot of communities of color today. So today we see a very large racial wealth gap, which is why it's so important to have equitable economic policies.[2]

Relatedly, when a lot of Black and Brown communities were forced into more urban areas because it was the only affordable place left, highways were constructed there to separate Black communities from white communities. These places experience a lot of transportation-related pollution and the consequential health outcomes from that pollution.[3] A lot of polluting facilities, whether oil and gas refineries or wastewater sites, are also located in those communities because there wasn't a lot of political power to resist the siting of the facilities there.

So then, come to the present day, historically redlined communities have very failing infrastructure, scarce economic opportunities, and countless environmental hazards as a result of the accumulation of a lot of really harmful, racist policies. All of this translates into a climate gap, where climate impacts affect people of color disproportionately and unequally. That is why our policies cannot be race neutral. We must be explicit about race because so many harmful policies are based on race. We can't just talk about the physical environment alone because it's always connected to economic opportunity, to pollution, to health outcomes. These issues don't exist in siloes, so they require comprehensive solutions.

On the comprehensive climate policy work she's advocating for at the state level

As the associate director of climate equity at the Greenlining Institute, I have the opportunity to advocate for truly intersectional climate resilience and equity policy at the statewide level. For instance, last year we put out the report *Making Equity Real in Climate Adaptation and Community Resilience Policies and Programs*, and part of my work is finding opportunities in legislation to incorporate our recommendations and train different policy makers and agency staff on what operationalizing equity can look like. We recognize that there's a strong commitment to advance equity in California, but we want to make sure that turns into an actual strategy and practice and that there's accountability. This report has been helpful in providing policymakers with a blueprint or roadmap on how you actually do that work, because we don't want our policymakers to be trying to figure that out on their own. It's important that they're learning what effective equity policy looks like from the communities who have a lot of firsthand experience and expertise in that.

More recently I've been working closely with my supervisor to develop a proposal for a statewide office on racial equity. Washington State and San Francisco City have racial equity offices, so we're using those as models to consider what it could look like for the State of California to have an office whose role would be to train agencies and departments on racial equity screening methods, to identify racial disparities across agencies and the work that they're doing, and to develop strategies to address those racial disparities. The idea is to really utilize the office to train folks across the government, so in case funding for the office were to run out, the institutional knowledge would still be there.

I'm also working with Amee Raval[4] of APEN on creating a mapping platform to help the State of California identify vulnerable populations. It's relatively easy to create a tool, but we want to make sure that there's a lot of ground-truthing that informs the tool and what indicators show up. When I use the term *ground-truthing*, I mean that when we're thinking about using data to inform policy, data does not convey the full picture. It often leaves out the human component. Unless we're actually speaking with people who are experiencing the impacts that policymakers are trying to address, listening to their stories and their lived experience, all of the stuff that's harder to put in numbers, then we're not getting the full picture or holding ourselves accountable to what communities might need. Ground-truthing is elevating and uplifting the shared experience and expertise of community members and acknowledging that, even if they're not policy experts, the very fact that they're going through these experiences means they know best what their community needs. If we're not connecting with them early on and often throughout the process, it's just a missed opportunity for us to be able to make truly impactful change.

For instance, there have been a lot of conversations recently about the importance of having more cooling centers available for folks without air-conditioning on hot days. However, we often see cooling centers built as rooms with air-conditioning; they don't have services like beds to rest on, or water, or activities for children. We saw one cooling center that you'd have to literally cross highways to get to if you were getting there by public transportation. If you're not aware of the nuance of how a community might use a cooling center, you might create one that's not usable.

On the power and limitations of policy

I want to preface this by saying that policy is *one* strategy. We need a suite of strategies to really make the type of systemic, transformative change that we need, and we all have our own parts to play in that. But for me, the reason why I love policy work is because of its proactive nature. When I was working in law, it felt more reactive; we were trying to fix something after it had gone wrong. Particularly with climate resilience work, it's so important that we're able to issue-spot challenges before they actually turn into real problems, and we're able to think about ways to address challenges before things get really bad.

One reason I really like doing policy work at the Greenlining Institute specifically is because we never stop at a bill passing; we're always tracking the implementation of the bill and analyzing whether it's meeting the goals that it set out to achieve or if there are any unintended consequences, because implementation is where all the real stuff happens. If we're not tracking its implementation, we really don't know what kind of impacts the policy is creating.

This approach also forces us to have an inside/outside strategy to our work. When we're working in Sacramento, we're focused on building leverage from within the system; we're attuned to any political context that could prevent good policy from moving forward, and we're actively building equity champions within the legislature so that they can advocate for our priorities. At the same time, we're working in coalition with community-based organizations across the state to put a lot of pressure on our decision makers from outside the system. Policy work allows us to exist in both of those spaces, doing campaign and external advocacy work in relationship with organizers on the ground while also building trust and understanding with decision makers so that we can move good policy forward.

There are many reasons why good policy isn't enough, though. Funding, of course, is always a big barrier. For instance, our vulnerability mapping platform was very well received, and we got buy-in for it, but until we get funding for the program, it will be impossible for us to implement this thing that is technically very easy to implement. And that program requires a relatively tiny amount of funding. Truly impactful climate resilience policy requires millions and billions of dollars. Additionally, many programs require that communities pay up front and get reimbursed later. Therefore, the communities that are most strapped for resources are effectively excluded from the very policies and programs that are meant to lift them up. These are the details that I didn't even think about until I started doing this work but end up making all of the difference.

On connecting the struggles for climate justice in California and Louisiana

I grew up in New Orleans. I was there for Hurricane Katrina, a lot of oil spills, and many other devastating hurricanes. While I didn't have the vocabulary to articulate what I was seeing at the time, I remember being a senior in high school and noting that about 60–75 percent of my classmates, the majority of whom were lower income students of color, weren't able to return after Katrina hit. I stayed in New Orleans for college, when the rebuilding was starting, and it was clear that the French Quarter and Bourbon Street were getting a lot of investment and were being rebuilt very quickly, while other parts of the city, in predominantly Black or Vietnamese communities, weren't getting a lot of resources or investments. Even today some of those neighborhoods feel kind of like ghost towns. It didn't make sense to me at the time, but it always stuck with me.

Today I live in Oakland. While New Orleans deals with hurricanes and Oakland is more impacted by wildfires and smoke, both communities face similar climate issues in terms of sea-level rise, flooding, and extreme heat. So much of the climate policy work that I do in California is applicable to Louisiana, as well. One day I hope to go back and do similar work there. I would like to give back to the city that has brought me so much joy and has been so foundational to my identity. If I'm able to help my hometown heal from so many environmental harms, as well as the legacies of the city's harmful, racist histories, I know it will help *me* heal, too.

RESILIENCE TOOL SPOTLIGHT
Inside/Outside Strategy

Inside/outside strategy, also called IOS, is a political and organizing strategy utilized to broadly shift public opinion and incite transformational social change. The strategy hinges upon growing the number of social-, racial-, and climate-justice champions embedded inside political institutions while building robust grassroots movements outside those institutions. On the inside, people in positions of power, such as political representatives, members of commissions and boards, and union and advocacy group leaders, work to funnel resources to the outside, champion the visions of the mass movement, and actualize meaningful climate action and justice-oriented legislation, projects, and more. Meanwhile, on the outside, activists organize and nurture movements, develop solutions grounded in the lived expertise of community members, engage in protest and direct action, and push insiders to act more boldly. Together they create synergy. For IOS to be successful, however, folks on the inside must remain committed to revolutionary change and accountable to the movements that they are representing, and grassroots movements need to be large, loud, and savvy enough to mobilize the public and maintain pressure on the folks who are supposed to serve them.

Among many valid critiques of inside/outside strategy, some argue that it is oxymoronic to seek liberation within structures designed to be oppressive; trying to do so will only expand or reify harmful institutions and water down revolutionary agendas. Champions of IOS, however, contend that good IOS can transform existing structures while building new ones. They reject the idea that our movements need to be confined to either political or grassroots spaces to make a significant difference, and instead embrace both/and thinking. They explain that, by maintaining a diligent analysis of power and focusing on building local projects deeply rooted in community, IOS can help us move toward governing ourselves and reclaiming control over food, water, art, work, and power, all the while making material differences in people's everyday lives.

Like all resilience tools featured in this book, inside/outside strategy may not be appropriate for every context or feel values-aligned for everyone. For folks with whom it resonates, however, IOS can be a profoundly powerful, transformative movement-building tool when wielded thoughtfully, and it's an excellent framework to keep in mind when engaging in any political or community-building work.

To get involved with inside-outside strategy in action:

- **Research prompts:** If elections are approaching, are there any campaigns underway that you're excited to support or engage with? Are there any commission, committee, council, or board positions open in your municipality that you might be interested in applying for? (If there's a climate action commission with openings, that could be a great place to start, but keep in mind that there are a plethora of areas where you can do fantastic work advancing intersectional climate resilience solutions.)

- **Journal prompt:** If you were to run for office, what would your campaign priorities be? Which communities/movements would you be deeply engaged with and accountable to? (Note that these are especially great prompts to consider together with pals and comrades.)
- **If you have the bandwidth and interest in getting more involved in local or regional politics:** Begin by getting a better sense of the different ways that you can get more involved in political spaces and what makes sense for you. If you're interested in joining a local or regional commission, committee, or board membership, programs like Urban Habitat's Boards and Commission Leadership Institute empower community members with the skills and confidence to apply and serve effectively. If you're interested in running for office, there are plentiful organizations that offer training and support, including EMILY's List for pro-choice Democratic women, Higher Heights for America for Black women, Victory Institute for queer and trans candidates, Run for Something for young progressives interested in local positions, and New American Leaders for first- and second-generation Americans. The list goes on. Running for office can be a daunting and time-consuming undertaking, so finding a community of learning and support can go a long way.
- **If you're excited to support a movement comrade or values-aligned activist or advocate running for office:** There are infinite ways to support their campaign. If you know them personally, ask what they might need. Many candidates need help with everyday activities like cooking, cleaning, and childcare while they're running for office. Bring your skills and passions to their campaign as a volunteer. Offer to host a fundraiser at your home. Gather pals to canvas door-to-door or phone bank. Donate to their campaigns and to the grassroots movements that will hold them accountable.

KAILEA FREDERICK

SHE/HER/HERS

"The return of the matriarchy is the possibility of moving back into balance"

Kailea Frederick is a multihyphenate powerhouse. When I think about the many different tools needed for the climate movement to win, I think of Kailea, because she is wielding about half of that toolbox. She dances between roles as a mother, organizer, facilitator, writer, creative, relative, and speaker at the intersections of climate justice, spiritual ecology, resilience thinking, and Indigenous sovereignty. Presently she is the coeditor for Loam and she recently transitioned from being a climate justice organizer with NDN Collective to becoming their community publisher. In 2022 she concluded her term on the Climate Action Commission of Petaluma, California.

With Loam, an organization that seeks to support and uplift compassionate, creative climate activists, Kailea has curated and edited an annual magazine, helped actualize their *Loam Listen* podcast, and coauthored and edited invaluable books like *Compassion in Crisis* and *Living through Liminality*. With NDN Collective, an Indigenous-led organization dedicated to building Indigenous power through organizing, activism, grantmaking, capacity building, and narrative transformation, Kailea has championed Indigenous climate solutions, strategized savvy campaigns, and edited groundbreaking texts like *Required Reading: Climate Justice, Adaptation, and Investing in Indigenous Power*. Finally, as a former member of the Petaluma Climate Action Commission, Kailea was instrumental in coauthoring and helping successfully pass the Petaluma Climate Emergency Framework, which will guide city policy around climate mitigation, adaptation, and justice. Whether she's seeding climate justice in the nonprofit space, creative industry, local government, or beyond, she leads with nuance, compassion, and wisdom well beyond her years earthside.

Content Notes: Settler colonization, sexual assault, and gendered violence

On her journey into the intersectional environmentalism movement

I was pulled into the environmental movement in 2014 while living in Maui. I didn't realize I was being initiated into activism at the time; I just thought I was standing up for my community. Since Hawaii has a year-round growing season, the islands became a testing site for companies like Monsanto to test out new chemical combinations.[1] They would test these potentially really poisonous chemicals near people's homes and schools, and we were finding traces of the chemicals in children's hair. I joined SHAKA (Sustainable Hawaiian Agriculture for the Keiki and the 'Aina), a people-led movement, to help try and place a moratorium on companies testing their chemicals on our lands until we could have our own scientists come in and ensure that it was safe.

Around the same time, the Mauna Kea movement was gaining public traction.[2] Both of these movements asked what it meant for the Hawaiian people to fight to protect their land. I feel so grateful that the beginning of my journey into intersectional environmentalism was grounded in the narrative of Native peoples fighting for their land.

On galvanizing action in local politics and the climate crisis in Petaluma

Today I split my time as the vice chair of Petaluma's Climate Commission and the coeditor of Loam.[3] I've served on the commission as a volunteer for the last couple of years. We're a policy advisory body on climate change for Petaluma's city council. It's an opportunity to educate our residents on the impacts of climate change, the ways that we need to adapt and mitigate, and what it all will mean for their day-to-day life. It's exhausting work, and it has asked me to shift how I work with people from different cultures and with different theories of change. But especially in this moment, when our country is having a very deep reckoning with not just climate change but also racial and social justice, it's so important that people who are truly committed to anti-racism have the opportunity to be in places where policies are shaped and implemented.

We recently passed the Petaluma Climate Emergency Framework, which is really exciting. And as any climate activist knows, it's also a little bit scary. It's almost like we're creating these pieces of legislation too late. I consider Petaluma to be on the front lines of climate change. Each year since 2017, we've dealt with consecutive fire seasons that have ravaged the area. The fire season has lengthened to the point that a major, out-of-control fire could now happen at any time of the year. The climate crisis creates conditions ripe for uncontrollable fire; the winds are stronger, and we've been dealing with a multiyear drought. On top of that, the land has been mismanaged for decades; California's ecosystems depend on controlled burns to thrive and stay healthy, and when they don't burn on a regular basis, they become overstocked with dry kindling, so all that's needed is a spark for a fire to become so massive and hot that it takes out entire towns and communities. It's the perfect combination for absolute devastation.

In 2017, within twenty-four hours of finding out that I was pregnant, the Tubbs Fire came through and burned down a portion of Santa Rosa, just twenty minutes from my own house. The following year, there was the Camp Fire in Paradise. Even though it was a few hours away from us, the fire was so big and the wind was so stagnant in the Bay Area that Petaluma was blanketed in smoke for weeks. It was traumatizing to be sitting under a blanket of smoke, in part because that smoke is heavily toxic, but also because we were so aware that the fire had taken people's lives, pets, and whole homes.

This past year was particularly hellacious. A lightning storm sparked hundreds of fires across the state. The smoke was unbearable, but on top of that, we had a heat wave with several consecutive days above 100°F, there were rolling power outages so we couldn't rely on air-conditioning to cool down, and there was nowhere safe to go because we were in the middle of a pandemic. There was one day when the smoke and the heat felt so suffocating that we went to Tomales Bay, where we located a pocket of relatively safe air on our air quality index map. There was a forest fire just upwind of the beach, and I have a photo of us swimming, trying to cool our bodies in the water, with plumes of smoke from the forest fire right behind us.

So the passage of the Petaluma Climate Emergency Framework is bittersweet, but I'm glad that we are doing the best that we can in the timescale that we have. The Climate Emergency Framework will be a lens for all city staff to view their work through, as well as a guiding document for the policy and large city plans that the city will be developing over the years. They're updating the General Plan right now, and it's really wonderful that they'll be able to ground that in the reality of climate change and incorporate tangible solutions that we outlined in the framework.

We've also recently invited youth to the table as voting members. At the very first meeting that we ever had, I started asking, "Where are the young people?" As a parent and in the last year of my twenties, I feel like I've aged out of the youth climate movement, but I came from and was trained in that movement. I am very aware of how capable young people are and how eager they are for training to be able to address these issues. We need them at the table because they aren't afraid to speak the truth, they're very aware of the timelines that we're dealing with, they're not driven by personal political gains, and they have strong moral compasses. I always want to uplift youth and empower them so that they feel like they have a foundation that they can continue building upon, no matter where life leads them.

On the importance of creative outlets and heart-centered expression

Separately, working as the coeditor with Loam has been a dream. Loam is a really tiny arts and environmental publication and publishing space founded by our creative director, Kate Weiner. It's a place where I get to bring all my creativity and imagination around how I want the new world to be built and all the perspectives that I want to be at the forefront. Kate and I consider our job to be shifting the cultural narrative around what it means to be alive on planet Earth today, what it means to be an activist. We're really dedicated to creating spaciousness in the movement.

Kate and I feel that, by offering space for people to express themselves freely, without creating a lot of stringent boundaries, we're able to receive more honesty and truth-telling. We hope that these expressions, in the form of writing, photography, podcasting, design, and illustration, are able to impact people's heart space and emotional fields. We haven't won in our movements by only being data driven and presenting the facts. We're trying to get people to consider new perspectives by seeing and feeling the humanity in what's being shared.

On rematriation as moving back into balance

Recently Kate and I have been discussing how to talk publicly about matriarchy. I was initially introduced to the concept of matriarchy through my own culture. I come from two different First Nation spaces that were traditionally matriarchal and still practice aspects of matriarchy, like following the lineage of clan systems through mothers. The return of the matriarchy is a reclamation of those societies that were traditionally matriarchal. I've also heard rematriation defined as a returning of the land, our language, our culture, our bodies, and our seeds to the sacred.

When I talk about matriarchy or rematriation, I also look at the ways that extractive industry is directly tied to both the pillaging and raping of the earth's body, alongside the pillaging, raping, and disappearance of the bodies of Native women and two-spirits who are within proximity of these types of industry, culture, and workspaces. When I went to high school in British Columbia, I lived directly off the **Highway of Tears.** It was so drilled into me that I should never, ever walk on that highway. There was such a fear of my body being taken because of my identity and social location.

The return of the matriarchy is the possibility of moving back into balance. It's not about women or femmes being in oppressive power. It's not about changing out who gets to be in charge. It's actually so much more about the values that we hold ourselves accountable to as a larger society. It's about valuing interconnectedness and the reprioritization of children and elders. I think often about how different the pandemic could have been if American culture actually prioritized children having an education and elders and folks who are immunocompromised or disabled not being exposed to Covid-19.

DOMINIQUE THOMAS

SHE/HER/HERS

"We don't all need to be doing the same work, but we do need to have similar goals and values"

Dominique Thomas is a Harlem, New York–based grassroots organizer, researcher, trainer, and Afrofuturist. I learned about her activism through Lil Milagro Henriquez, who mentioned that Dominique absolutely brought the house down while speaking at a Mycelium Youth Network conference. After I had the honor of conversing with Dominique, it was easy to see why. Not only does she wax poetic about organizing strategy with magnetism and ease, but she coaxes people to dream bigger and truly believe in radical possibilities. Dominique honed her expertise in transformative relationship-building and organizing while cochairing the New York City chapter of the Black Youth Project 100 (BYP100), a national, member-based collective of Black youth activists dedicated to creating justice and freedom for all Black people through direct-action organizing, advocacy, and political education. After several years of working in healthcare by day and mobilizing community after hours, Dominique began organizing full-time with longstanding climate action organization 350 as their New York and Mid-Atlantic regional organizer. While building statewide coalitions and supporting more than a dozen 350 chapters across the region, she identified a need to better support frontline activists. She thus cocreated and launched the Climate League, a racial justice–centered climate training program through 350 for BIPOC folks to deepen organizing and campaigning skills and be in community with one another.

When Dominique and I spoke, she was working as a training manager with Climate Advocacy Lab, a nonprofit on a mission to build grassroots power and win through evidence-based advocacy and more effective public engagement campaigns. Since then, Dominique has joined Industrious Labs, a climate startup dedicated to decarbonizing the industrial sector, advancing a circular economy, and building a just and equitable transition. As a partner and field building director, Dominique is using her organizing and visionary prowess to build transformational relationships and power with frontline and worker communities most impacted by extractive industry.

Content Notes: Structural and institutional racism, racialized policing, slavery

On her journey into intersectional climate organizing

After graduating from college and moving to New York City, I quickly realized that the work I was doing in clinical trial research wasn't helping the communities that I wanted to be working with. Around the same time, I started organizing with Black Youth Project 100, a national organization operating from a Black, queer, feminist lens and fighting for social, economic, educational, and political freedom. We were doing a lot of organizing around H.O.M.E. (Housing over Monitoring and Exclusions), a campaign to end Permanent Exclusion, a policy that calls for anyone who is arrested to be evicted and permanently banned from public housing.[1] We were part of the Swipe It Forward coalition and would swipe Black and Brown people into train stations around the city for free as a means to disrupt **broken windows policing**, which is utilized to arrest folks who jump turnstiles and then permanently exclude them from housing.

For example, if I was arrested while jumping a turnstile and I lived with my mom in public housing, I would no longer be allowed to live with her or step foot on her property. My mom would have to sign a document stating that she wouldn't allow me to live there, and her house could be monitored at any time to check that I wasn't there. It's a malicious, heinous system that not only creates a pipeline of folks into the carceral system but also surveils them, their families, and their loved ones.

As we organized around the H.O.M.E. campaign, we naturally started having a lot of conversations about what *home* means. People living in New York City public housing deal with lead paint and a lack of heating and running water. There's no air-conditioning offered even though New York City summers are extremely hot and humid. There are several waste treatment facilities, highways, incinerators, and toxic waste dumps located within our communities, and parts of Washington Heights and the Bronx have the highest asthma morbidity rates in the country.[2] For me, it begged the question, *What does home mean when you're living in a place designed to treat you without dignity?*

I had heard about global warming in school, but I really started connecting the dots between racial justice, environmental justice, and climate justice. I knew that I wanted to be organizing more than I wanted to be doing research, so I took a job as a Mid-Atlantic regional organizer with 350, known as a **Big Green.** Since many Big Greens have white, older bases,[3] in part because they're funded by white philanthropy, it felt important to highlight that many of our BIPOC siblings have been dealing with environmental racism for decades, and climate impacts like sea-level rise and extreme weather events will be experienced differently by folks who are already marginalized because of imperialist, white-supremacist culture and the bodies that they live in. Whether I was building relationships with local groups, giving presentations in middle schools, or providing political education in climate trainings, it felt really crucial to uplift the stories and lived experiences of folks on the front lines and to connect the climate crisis to the interlocking crises that are happening.

When we talk about the climate crisis, we also must talk about racial capitalism, economic injustices, and white-supremacist culture. And we need to not just talk about those things, but we also must actively be dismantling those systems. I want us to be more strategic in both calling out the root causes of these issues and also thinking about how we can be more thoughtful in building up the power necessary to dismantle those systems. Because if we work from a place of interconnection and intersectionality, we will simultaneously defeat the climate crisis and white supremacy and create equity for all people.

Some of that work is already happening, but perhaps not yet at the scale required. There's still a big fracture between environmental justice work and more mainstream environmentalism. In my job as the training manager of the Climate Advocacy Lab, I'm thinking a lot about what it could look like to bridge and marry the fractured movements so that we can build the people-powered movement that we need to win.[4] We don't all need to be doing the same work, but we do need to have similar goals and values.

On the importance of relationship-building in organizing

As an organizer, my job is to make myself obsolete and unnecessary in those spaces. I don't believe in a hyperfocus on super charismatic movement leaders. Instead, my focus is on building leaderful movements. As Ella Baker said, "Strong people don't need strong leaders." If we're organizing effectively, we're retaining the people that we recruit because they've found a sense of identity and community in the work. At each rung across the ladder of engagement, we're continuously developing their leadership skills. I'm not here to tell them what empowerment or agency looks like, but I do have a set of transformational relationship-building tools that I can utilize to help invoke what power and agency looks like for them, in their own life. I'm working to awaken folks to determine what success, solutions, and resilience look like in their own communities and lives. If we do that foundational work of empowering community, we'll win our campaigns. Maybe not in six months or a year, but eventually. Every win is an outcome of transformational relationship-building and power-building.

On Afrofuturism and the power of a vision practice

Afrofuturism is about reclaiming a narrative of visions of the future that don't include us or that we've been erased from. It's about using science fiction and technoculture—spaces that typically don't include images or stories of people that look like us—to envision a different future and to develop a theory of change for collective liberation.

I think a lot about Harriet Tubman, a Black woman who was enslaved during the 1800s, who had a severe, unpredictable, and incapacitating disability and was still fiercely committed to the work of liberation. She had such a clear vision of what liberation looked like for herself and for her family, and she was able to organize to empower other people to have that same vision. While she was considered an outlaw, on the run with slavecatchers after her, she still had people staring in the same direction as her, with the same vision. She was able to invoke and empower in them their agency to see and actualize something better for themselves. No matter what folks said or did to her body, they couldn't take her mind or the liberation that she'd claimed for herself in her mind. And she was able to quite literally bring other people on this vision with her without a map, internet, or GPS by following her gut and the path of the North Star.

In my mind's eye, I can see the future that I want for people who look like me. And if I can see it in my mind, you can't tell me that it isn't real. Once that vision is activated within my mind and the minds of people around me, nobody can take that away. When things get hard, I can always go back to that vision of people being happy, thriving, and free. That's so powerful. If people didn't have a vision of a life that was better than what they were being given, we wouldn't have had the civil rights movement. It's important to remember that people created this world that we live in, and if people can create the cruelty and horribleness of these systems, people can also dismantle them and dream up and build something better. So for me, that's the center of organizing and how we'll be able to win on anything.

RESILIENCE TOOL SPOTLIGHT

Cross-Movement Building

Cross-movement building is a time-honored strategy utilized by distinct (though often overlapping) groups who come together, typically with the understanding that they share the same goals or are impacted by the same grievances. Recognizing that both their oppression and their liberation are linked with one another's, they work together to organize, share resources, and form a united front.

At this point in the book, it should be supremely evident that climate action intersects *all* movements for collective liberation and justice. Climate justice is foundational to economic justice, racial justice, disability justice, reproductive justice, gender justice, environmental justice, health justice, immigrant justice, Indigenous sovereignty, queer and trans liberation, fat liberation, Black liberation, and on and on. And vice versa.

Since the climate crisis exacerbates the injustices that marginalized communities already experience, and since the climate crisis is produced by the same systems of white supremacy, settler colonization, imperialism, growth-at-all-costs capitalism, ableism, and cis-hetero patriarchy at the root of other ecological and social crises, cross-movement solidarity is the only way forward.

I hope that this gives you permission to dream about what climate action can look like beyond designated "climate spaces." The climate movement needs more folks weaving connections and building solidarity between and across people, organizations, and movements. It needs more people mobilizing their comrades for climate action within the spaces where they're already active, in the communities where they're already deeply connected. The climate movement needs more humans who are committed to the fight for the long haul because they are doing work that they're passionate about and are sustained by the relationships that they create along the way.

If none of the action items presented in this book have resonated with you thus far, that's okay. What's important is that you find ways to get involved that feel joyful, meaningful, and nourishing for you (at least most of the time, because this work will likely feel tedious, challenging, overwhelming, and uncomfortable some of the time, too, and that's also okay). Because the fight for a livable planet for all will be lifelong and will require all of us to win.

To help identify where and how you may feel called to act in service of the climate and the constellation of movements for collective liberation and justice:

- **Consider where you would like to call your movement home(s):** The fight for a livable planet can and should be happening across all social change spaces, so know that you're not limited to working on climate issues from within the so-called climate space. When thinking about which community group, organization, or larger movement(s) that you'd like to be most rooted in, consider which issues or identities are closest to your heart. Research the work that's already happening in your community, and think about where you might feel most supported, safe, challenged, and inspired, who you'd like to surround yourself by, and where you believe your perspective is needed and gifts will be appreciated.

- **Consider how you can bring your superpowers to the movement:** Marine biologist, writer, policy expert, and climate activist Dr. Ayana Elizabeth Johnson created the brilliant Climate Action Venn Diagram, which asks people to consider what they are good at, what brings them joy, and what work needs doing. Environmentalist and drag queen Pattie Gonia (@pattiegonia) also urges folks to consider the identities that they hold and the communities that they're a part of. The sweet spot where all those spheres overlap is where you should begin.
- **Consider the role(s) that you'd like to embody:** In the 1970s organizer, journalist, politician, and social change activist Bill Moyer coined the Four Roles of Activism in his Movement Action Plan: the rebel, the reformer, the change agent, and the citizen (activist and writer George Lakey updated these names to the rebel, the advocate, the organizer, and the helper, with Moyer's permission). Many activists, scholars, and changemakers have since expanded upon and brought nuance to the concept of the different roles that comprise a movement organizer. Check out the "What Is My Role in the Climate Movement" framework by Joycelyn Longdon on her Instagram page @climateincolour and the "Callings & Roles for Collective Liberation" framework put forth by Slow Factory (@theslowfactory). Let go of misconceptions of the roles an activist should inhabit, and think about which roles you identify with or gravitate toward.
- **Consider how you can collaborate with and support others:** Ultimately our work is so much stronger together, and we will need a whole lot of us working in unity to win. So think about how you and your comrades can build meaningful relationships with other individuals,

organizations, and movements. Then start considering how you might collaborate strategically, lean on one another, and stand with them in solidarity when the moment calls for it.

Please find more essays, resilience tools, and resources for continued learning at www.climateresilienceproject.org.

CONCLUSION

When I first began working on this project three years ago, I was just as interested in exploring the *how* as the *what* when it came to climate resilience strategies. It felt like so many so-called climate solutions that were promising on the surface were failing to deliver—or even causing immense harm—in practice. Things often seemed to go awry during the implementation process. The incredible leaders featured in this book, however, are enacting tangible, positive, and lasting changes in their communities while taking enormous care to prevent any possible unintended consequences of their actions. So during the interview process, we spent copious amounts of time discussing their approach, intention, and lessons learned along the way.

Many dozens of interviews later, and it's clear that there's no singular right way to engage in climate action. However, several common questions emerged that this book's contributors tend to ask themselves upon reaching a fork in the trail or in moments of doubt. So as you newly embark on your climate action journey or deepen the work that you're already engaged with, I'm leaving you with a handful of these questions in the hope that you can return to them when you're in need of gentle inspiration and guidance:

Who are my people, and how can I/we organize with them?

Who is already doing the work that I'm/we're interested in? What kinds of support are they calling for?

Who benefits from this work and who does not?

How are the folks who are most impacted being centered? Which perspectives are getting left out? How can I/we remove barriers or call folks in to help ensure their perspectives are included?

How can I/we balance the urgent nature of this work with the slow, intentional tempo required to nurture relationships and build trust?

> How can I/we integrate joy, rest, connection, healing, and fun directly into this work? How can I/we best pace myself/ourselves to stay engaged for the long haul, knowing that this work will likely continue beyond our lifetime?
>
> How can I/we regularly practice the world that I/we want to live in?
>
> What is the next right thing?
>
> What can I/we learn from the more-than-human world about how to best effect change? How can I/we be like mycelium, allowing an entire fungal network to develop from a single spore? Or like a raindrop, catalyzing a cascade of ripples upon reaching the water's surface? How can I/we mimic a perennial plant, embracing rest and hibernation in some seasons and growth in others? How might I/we move like water to erode resistance and expand the flow?
>
> Finally, what can I/we learn from the generations before us, and particularly those who lived closest to the land? How can I/we honor their experiences and humble myself/ourselves to their wisdom?

I'm so grateful that you joined me/us in this journey. I can't wait to cocreate a climate revolution with you!

GRATITUDE

My deepest, heartfelt thank-yous to:

All of the brilliant contributors. Thank you for generously sharing your wisdom and experiences, trusting me with your words, and championing a beautiful, just, and viable future for us all.

Larissa Melo Pienkowski, for being the best agent and advocate a writer could ask for. I'm endlessly grateful for your tireless, joyful support and guidance.

The North Atlantic Books team, and especially Gillian Hamel, for believing in and nurturing this project.

Raquel Pidal, for your keen eye and care.

Kait Heacock, for helping to uplift this work.

Mentors like Coleen Fox, Khalid Kadir, and many more, who have shifted my worldviews and challenged me not to settle for harmful climate "solutions."

My political homes, like Solidaire, Justice Funders, and Movement Generation, for fostering awe-inspiring communities and radical possibilities. So much of this work has been informed by the collective vision, love, and strategy of your ecosystems.

Ayana Elizabeth Johnson, Katharine K. Wilkinson, and countless women, nonbinary, gender-expansive, and two-spirit climate leaders, for ushering in this beautiful feminist climate renaissance.

Elders, ancestors, and more-than-human kin, who have been illuminating paths to community resilience for generations.

My beloveds, and especially Kyle, for cheerleading me every step of the way. Thank you for your patience, humor, and buoying support.

GLOSSARY

The definitions here aim to provide deeper understanding, context, and nuance around some of the terms in this text. Please note that language changes quickly and some of these terms and definitions may be dated by the time that you're reading this.

Afrofuturism A term first coined by scholar Mark Dery, who defined Afrofuturism as speculative fiction that utilizes African diaspora themes and issues and then places them within technoculture worlds, blending elements of science fiction, fantasy, and African diaspora culture to reenvision alternate futures, presents, and pasts where African-descended people play a central role in the creation of that world. Today Afrofuturism can be understood as a wide-ranging social, political, and artistic movement and a powerful tool to envision and create more just worlds.

Anthropocene A term coined in the 1980s and popularized by atmospheric chemist Paul Crutzen in 2000 to describe the time during which humans have had a substantial impact on the planet's physical, chemical, and biological systems, and particularly in driving an era of mass extinction. The term has been critiqued for pointing the finger at humanity rather than settler colonialism, white supremacy, imperialism, capitalism, etc., more specifically, and for contributing to the false human/nature binary, but it can be a helpful framework for describing this unique moment in time.

Arctic amplification The phenomenon of temperatures in the Arctic increasing about twice as fast as in the midlatitudes. When sea ice melts, sunlight that was previously reflected by ice's bright surface becomes absorbed by the dark ocean, therefore amplifying the warming process.

Big Green A moniker for the largest environmental organizations in the so-called United States such as Greenpeace, Sierra Club, the Nature Conservancy, and the Environmental Defense Fund.

Bodymind A term used by many disability scholars to emphasize the interdependence and inseparability of the body and mind and to challenge mind-body dualism.

Broken windows policing A model of policing first introduced in 1982 by James Q. Wilson and George Kelling, who argued that disorder (e.g., loitering, public drinking, jaywalking, spitting, biking on the sidewalk, and fare evasion) leads to increased fear and withdrawal from residents, which then allows more serious crime to occur. When Rudy Giuliani was elected New York City's mayor in 1993, he put the theory into practice, leading to overpolicing and criminalization of communities of color, as well as excessive use of force to address relatively harmless situations.

Brownfield site A term used by the real estate industry to describe lots that are characterized by the potential presence of a hazardous contaminant. Brownfields are generally the by-product of past industrial, commercial, or agricultural uses involving toxic products. Due to a long history of discriminatory zoning policies, which allowed hazardous industries to operate in low-income communities of color, there's a well-documented correlation between race, class, and brownfield sites.

Clean energy A term that refers specifically to sources of energy that don't emit greenhouse gases. It is not the same as renewable energy, as clean energy can include nuclear energy, natural gas with carbon capture and storage, and other nonrenewable sources of energy that don't emit carbon dioxide.

Community-supported agriculture (CSA) A farming model designed to allow farmers to receive advance working capital to invest in their growing season, gain some financial security, earn better crop prices, and spread both the risks and benefits of the farm among community members. In a traditional CSA model, community members buy a share of the farm's production before each growing season and receive periodic distributions of the farm's bounty throughout the growing season.

Critical race theory A framework and movement first developed by legal scholars in the 1970s and 1980s who were calling for legal approaches that take into consideration race and racism as a nexus of American life, not an aberration that can be easily corrected by law. Law professor Kimberlé Crenshaw, widely credited with coining the term, describes critical race theory as more of a verb than a noun: “It is a way of seeing, attending to, accounting for, tracing, and analyzing the ways that race is produced, the ways that racial inequality is facilitated, and the ways that our history has created these inequalities that now can be almost effortlessly reproduced unless we attend to the existence of these inequalities” (Jacey Fortin, “Critical Race Theory: A Brief History,” *New York Times*, November 8, 2021). Law professor Mari Matsuda, also an early developer of critical race theory, says in the same *Times* article, “For me, critical race theory is a method that takes the lived experience of racism seriously, using history and social reality to explain how racism operates in American law and culture, toward the end of eliminating the harmful effects of racism and bringing about a just and healthy world for all.”

Deep decarbonization A long-term strategy to gradually eliminate all carbon-emitting fuels from energy systems across sectors. The idea has been popularized by the Deep Decarbonization Pathways initiative, a collaboration of leading research teams across the world who are working to propose realistic pathways to reach carbon-neutral emissions by 2050. The initiative, led by the Institute for Sustainable Development and International Relations, has largely focused on informing science and policy at regional, national, and international scales.

Deep ecology An environmental philosophy and social movement rooted in the belief that humans must recognize the inherent value of nature rather than valuing nature narrowly for its utility to humans. The phrase was introduced by Norwegian philosopher Arne Næss and American environmentalist George Sessions in 1972.

Distributed energy Electricity that is generated at or near where it will be used. Distributed energy generation systems are generally relatively small-scale and modular, such as rooftop solar units, small wind turbines, emergency backup power generators, or biomass combustion, and can either serve a single structure or be part of a microgrid. Distributed energy can help ensure that important community buildings, like resilience hubs and hospitals, maintain electricity even when centralized power fails.

Energy burden The percentage of household income that goes toward energy costs, including electricity, heating, and transportation. According to the US Census Bureau, the average energy burden for low-income households is three times higher than for non-low-income households, a statistic that is further exacerbated by race and geography.

Food desert Geographic areas that have few or no convenient options for securing affordable and nutritious foods. Researchers typically identify food deserts by the number of grocery markets in a given area, the distance people have to travel to the nearest full-service store, average household income, availability of public transportation or vehicle availability, etc. It's an incomplete framework, as it doesn't accurately reflect the numerous and complex reasons why culturally appropriate and nutritious food may be out of reach, but it can help flag areas where food access is particularly limited.

Frontline and fenceline communities Communities that experience the first and worst consequences of climate change are often referred to as frontline communities. Neighborhoods that are situated next to industrial or military operations and are exposed to inordinate levels of hazardous materials, emissions, and environmental degradation, along with heightened risk of chemical explosions, are often referred to as fenceline communities. These communities often overlap.

Global North and South The terms Global South and Global North are used as politico-economic meta categories. The Global South generally encompasses the regions of Latin America, Africa, Oceania, and parts of Asia, while the Global North generally refers to North America, Europe, Australia, New Zealand, and certain parts of Asia. While the North/South binary is a tremendous oversimplification of a deeply complex and nuanced world system, it can be helpful shorthand when discussing power dynamics and broad trends.

Great Pacific Garbage Patch A massive area in the northern Pacific Ocean where spinning debris is drawn in and trapped by the North Pacific Subtropical Gyre. A tremendous amount of debris (mostly tiny bits of plastic) has accumulated over time, and it has since become emblematic of the world's litter, microplastic, and waste-production crises.

Greenwashing The practice of organizations exaggerating or lying to persuade the public that their products, practices, or policies are environmentally friendly. Rather than taking action to make meaningful changes, organizations often focus on marketing spin to appease public concern.

Healing justice A term and movement first coined by the Atlanta, Georgia–based Kindred Southern Healing Justice Collective in 2007. It refers to the reclamation of ancestral healing practices and the development of new, more inclusive practices in order to holistically address widespread generational trauma from systemic violence and oppression.

Highest and best use A concept developed by early economists that remains a fundamental principle of real estate appraisal. The Appraisal Institute defines highest and best use on their website as "the reasonably probable and legal use of vacant land or an improved property that is physically possible, appropriately supported, financially feasible, and that results in the highest value. The four criteria the highest and best use must meet are legal permissibility, physical possibility, financial feasibility, and maximum productivity." For instance, a highest and best use analysis of a contaminated lot may find that it is legally permissible and physically possible to develop apartment buildings on the lot, but it would cost so much to remediate the land that its value as currently used is deemed higher.

Highway of Tears An infamous, remote, 450-mile stretch of Highway 16 in British Columbia where many Indigenous women, girls, and two-spirits have been reported missing or murdered in the last several decades. The highway bisects several First Nations communities, and the region lacks adequate public transit, so many people are forced to hitchhike to get around. The Highway of Tears has become symbolic of the larger Murdered and Missing Indigenous Women, Girls, and Two-Spirit (MMIWG2S) movement.

Intersectionality A term coined by law professor and scholar Kimberlé Crenshaw in a 1989 paper to help explain the bias and violence uniquely experienced by Black women. She intended for the term to be a lens through which to assess where and how power collides, interlocks, and intersects.

Just Transition The concept was introduced by labor unions and environmental justice groups who recognized both the urgency of phasing out harmful and extractive industries and the importance of providing a fair pathway for workers to transition to other good jobs. According to the Just Transition Alliance's website, "'Just Transition' is a principle, a process, and a practice. The principle of just transition is that a healthy economy and a clean environment can and should coexist. The process for achieving this vision should be a fair one that should not cost workers or community residents their health, environment, jobs, or economic assets." The Climate Justice Alliance elaborates on their website, "Just Transition describes both where we are going and how we get there."

Liberation theology A social, political, and religious movement and praxis that seeks to understand Christianity through the lived experience of oppressed people and apply Christianity in service to the liberation of oppressed communities. Liberation theology formally emerged in Latin America and Black churches in the so-called United States in the 1960s. James Cone's 1969 book *Black Theology and Black Power* illuminated and popularized a vision for liberation theology that centered the humanity and liberation of Black people and challenged theological paradigms rooted in white supremacy.

Limited-equity cooperative housing A housing model in which residents purchase a development share, rather than an individual unit, and commit to resell their share at a predetermined price, therefore preserving housing affordability and limiting what residents can gain when they sell their shares.

Low-road contractors A slang term for contractors who repeatedly violate wage, safety, and discrimination laws. Low-road contracting tends to proliferate in the rebuilding period following disasters, particularly as potential contractors compete to have the lowest bid for government-funded projects, and undocumented workers are particularly impacted by wage theft and unsafe and inhumane working conditions during this time.

Man camps The temporary housing facilities situated by resource-extraction projects for predominantly male workers. There is a well-documented, significant link between these sites and a rise in gendered violence, and in particular the phenomenon of Missing and Murdered Indigenous Women, Girls, and Two-Spirits (MMIWG2S).

More-than-human A phrase coined by ecologist and philosopher David Abram in 1996 to refer to earthly nature in a way that both challenges human-centered worldviews and emphasizes that humans are part of, not separate from, nature.

Peak oil The hypothetical moment at which global production of oil is the highest it has ever been or will ever be. Peak oil theory was proposed in 1956 by geophysicist Marion King Hubbert, a researcher for the Shell Oil Company, who stated that oil production would follow a bell-shaped curve, declining as the finite resource became depleted. Over the last several decades, peak oil advocates have warned that rapidly diminishing oil supplies could spike prices and threaten political and economic stability. However, with oil production innovations like hydraulic fracturing, enhanced oil recovery, and horizontal drilling, energy analysts are now generally more concerned with oil demand than physical limits on production.

Permafrost Ground below the Earth's surface that has been continuously frozen for several years—typically hundreds or thousands of years. When frozen, it's harder than concrete. When permafrost melts, infrastructure (like roads, homes, and pipelines) collapses, communities become unstable, landfills begin leaking waste and toxic materials into nearby waterways, greenhouse gases are released, ancient bacteria and viruses thaw, etc. The melting of permafrost presents a host of grave issues for the immediate community and the world at large.

Popular education A for-the-people, by-the-people approach to education grounded in political and class struggle and social transformation. The concept emerged from Latin America in the early 1900s.

QTBIPOC An acronym that stands for Queer, Trans, Black, Indigenous, and People of Color. The term developed to highlight and center the specific needs and experiences of Black, Indigenous, and people of color within the LGBTQIA2S+ community.

Reproductive justice A term and movement coined by a group of Black women in Chicago in June 1994 who agreed that the women's rights movement was not adequately defending the needs of women of color and other marginalized women, nonbinary, intersex, and trans people. The reproductive justice movement was founded as an intersectional, access-focused alternative to reproductive rights advocacy, which narrowly focused on abortion choice and centered cisgender, white women. A few years later, SisterSong Women of Color Reproductive Justice Collective was formed to create a national and multiethnic reproductive justice movement. Reproductive justice, as defined by SisterSong on their website, refers to "the human right to maintain personal bodily autonomy, have children, not have children, and parent the children we have in safe and sustainable communities."

Soil sponge The living matrix of soil and underground organisms, such as plants, fungi, bacteria, worms, and insects, that together form a porous, well-aggregated, and sponge-like system that can retain its structural integrity when wet, thereby better capturing, storing, and filtering water, holding landscapes in place, and providing nutrients for the local food chain. The idea was introduced by Australian microbiologist and climatologist Walter Jehne in his 2017 paper "Regenerate Earth" for Healthy Soils Australia, connecting the concepts of soil carbon with a restored water cycle. Didi Pershouse has popularized the term as a way to move away from narrow narratives around soil carbon to discuss soil systems more holistically and unify action around soil health.

Solidarity economy A post-capitalist framework, movement, and broad set of practices that emerged from social movements in Latin America and Europe in the 1990s. The solidarity economy is distinguished by its prioritization of people and the planet over endless profit and growth. Solidarity economy practices are grounded in principles such as participatory democracy, cooperative or public ownership, equity, pluralism (i.e., there's no one-size-fits-all approach), and respect for the earth. A solidarity economy ecosystem might include community land trusts, cooperative housing, time banks, community fridges, barter networks, nonextractive lending, credit unions, worker-owned media, and so much more.

Systems thinking A concept first formally developed by MIT professor Jay Forrester in 1956 in response to traditional forms of scientific analysis in Western academia, which tended to focus on isolated, individual pieces of a system. Forrester, understanding that no pieces of a system are truly ever isolated, encouraged folks to look at the whole picture and work to understand relationships between system components. Donella Meadows expanded upon the systems-thinking framework in her book *Thinking in Systems* and other scholarship. Donella's work focused on helping folks get to the root of complex environmental and social problems rather than attempting (and failing) to address the symptoms.

Turtle Island A name often used by some Indigenous peoples to refer to the body of land now commonly known as North and Central America. The name comes from Indigenous oral histories that tell the story of a turtle that holds the world on its back.

NOTES

Introduction

1 When "so-called" is used to precede place names throughout this text, my intention is to emphasize that these are first and foremost Native lands and the borders and names imposed upon them by settler colonizers are not universally recognized.

2 IPCC, *Climate Change 2021: The Physical Science Basis. Contribution of Working Group I to the Sixth Assessment Report of the Intergovernmental Panel on Climate Change* (Cambridge, UK: Cambridge University Press, 2021).

3 Institute for Economics and Peace, *Ecological Threat Register*, September 9, 2020, www.economicsandpeace.org/wp-content/uploads/2020/09/ETR_2020_web-1.pdf.

4 Jason Hickel, *Less Is More* (London: Penguin Random House UK, 2021).

5 There are many cisgender men doing incredible climate work, but white, wealthy, abled cisgender men have been disproportionately centered in climate conversations for the past few decades, despite generally being the least impacted by the climate crisis. Therefore, I was intentional in highlighting perspectives that haven't historically been as centered and are more experienced with dealing with the impacts of the climate crisis firsthand.

6 Rebecca Solnit, "Big Oil Coined 'Carbon Footprints' to Blame Us for Their Greed. Keep Them on the Hook," *Guardian,* August 23, 2021, www.theguardian.com/commentisfree/2021/aug/23/big-oil-coined-carbon-footprints-to-blame-us-for-their-greed-keep-them-on-the-hook.

Reverend Mariama White-Hammond

1 In looking at some of the conservation movement's most impactful leaders, Gifford Pinchot served on the International Eugenics Congress and American

Eugenics Society; Madison Grant wrote the white-supremacist book *The Passing of the Great Race, or The Racial Basis of European History* that Hitler later referred to as his Bible; John Muir, who founded the Sierra Club, regularly used racist and derogatory language to describe Indigenous and Black people and conceived of "wilderness" as a place free of the Indigenous who lived there; and on and on. Conservation's racist roots have profoundly influenced modern-day environmentalism, from perpetuating the blaming of environmental ills on overpopulation to forcing Indigenous peoples off their homelands in the name of wildlife and biodiversity protection.

2 Countless scientists and climate experts agree that 350 parts per million of global atmospheric carbon dioxide is the level of carbon dioxide considered "safe" to prevent major climate impacts. For context, during the Industrial Revolution the global average amount of atmospheric carbon dioxide was about 280 ppm, and in May 2022 atmospheric carbon dioxide was averaging 420 parts per million.

Ruth Miller

1 The first major oil discovery in Alaska was made in 1957, and two years later Alaska was proclaimed a state of the so-called United States. The oil and gas industry has since extracted tens of billions of barrels of oil and natural gas from Alaska.

2 Beginning in 1867, after Alaska was acquired as a territory by the so-called United States, and enduring for more than a century afterward, Indigenous Alaskan children were forced into a violent and inhumane boarding school system with the goal of cultural assimilation. Boarding schools were overseen by the federal Bureau of Indian Affairs, private churches, and later the State of Alaska.

Niria Alicia

1 Richard Schiffman, "An Insurance Policy for Climate Change? How Seed Banks Are Protecting the Future of Food," *Yes!* magazine, December 10, 2014, www.yesmagazine.org/climate/2014/12/10/insurance-policy-climate-change-seed-banks-future-of-food.

2 Doug Pibel, "Why You Should Be Saving Seeds (Even If You Don't Have a Garden)," *Yes!* magazine, April 25, 2017, www.yesmagazine.org/issue/science/2017/04/25/why-you-should-be-saving-seeds-even-if-you-dont-have-a-garden.

3 North Carolina State University, "Irish Potato Famine Disease Came from South America," Science Daily, March 7, 2007, www.sciencedaily.com/releases/2007/03/070302082530.htm.

Morgan Curtis

1 Joanna Macy, an author, environmental activist, Buddhist scholar, and deep ecologist, developed the "Three Stories of Our Time" model to help folks make sense of the world and clarify a path forward. One of those stories, the Great Unraveling, illuminates the destruction and collapse of biological, ecological, economic, and social systems.

2 Morgan redistributes funds to the Movement for Black Lives, the Northeast Farmers of Color Land Trust, and Decolonizing Wealth Project's #Case4Reparations.

Casey Camp-Horinek

1 The Dawes Act, passed in 1887, broke up reservation land into small allotments that individuals were supposed to live and farm on. It was a tactic to assimilate Indigenous peoples into white American culture and reinforce the ideal of individualism over collectivism.

2 The concept of blood quantum was first developed by the Colony of Virginia in 1705 to reduce the civil rights of Indigenous people. It was initially predicated upon the darkness of a person's skin but continued to evolve in a way that would most benefit the US government. In 1934 the Bureau of Indian Affairs instituted the calculation of blood quantum and official "Certified Degree of Indian Blood" cards by using tribal documents to attempt to trace back the original enrollees of a tribe. Native American studies scholar Elizabeth Rule explains that blood quantum "emerged as a way to measure 'Indian-ness' through a construct of race. So that over time,

Indians would literally breed themselves out and rid the federal government of their legal duties to uphold treaty obligations." (Kat Chow, "So What Exactly Is 'Blood Quantum'?," NPR, February 9, 2018, www.npr.org/sections/codeswitch/2018/02/09/583987261/so-what-exactly-is-blood-quantum.)

3 Hydraulic fracking has led to companies "disposing" of wastewater—a combination of extremely salty brine and chemicals—by injecting it into a deep layer of porous rock. The injection process changes the pressure within rock pores, triggering frequent earthquakes.

4 Casey Camp-Horinek and I spoke during the devastating cold wave that brought record freezing conditions to Texas, Oklahoma, and beyond in February 2020.

5 The Trail of Broken Treaties was a political protest led by American Indian Movement activists calling for the legal recognition of treaties, the restoration of the treaty-making process, the return of stolen Native land, and the reform of federal-tribal relations.

6 Of the 1,322 Superfund sites (highly polluted locations designated by significant hazardous material contamination) in the so-called United States, 532 are located upon Indian trust land, an astonishingly disproportionate figure given that Indian trust land makes up about 2.3 percent of the country's landmass. (Terri Hansen, "Kill the Land, Kill the People," Indian Country Today, updated September 13, 2018, https://indiancountrytoday.com/archive/kill-the-land-kill-the-people-there-are-532-superfund-sites-in-indian-country.)

7 Hickel, *Less Is More*, 66.

8 Gopal Dayaneni's lecture was part of Movement Generation's online course series "ReKINdle 2021: An MG Study & Strategy Series" that took place in fall 2021.

Victoria Montaño

1 The displacement and forced cultural assimilation of Lisjan Ohlone people began with Spanish soldiers and missionaries in the late 1700s. After California became part of the so-called United States in 1850, the state government sanctioned mass genocide of the Ohlone people in an attempt to eliminate

Indigenous Californians. By 1852 just 10 percent of the precolonial era population remained.

2 Corrina Gould explains, "Shellmounds are created by my ancestors as ceremonial places and as burial sites." As recently as 1909, there were still 425 shellmounds around the shores of the San Francisco Bay. Today four of those shellmounds are still visible. Corrina, Johnella, and others have been working for decades to protect what's left and restore sacred sites that have been built over.

3 Victoria notes that medicine like California white sage has recently become trendy and therefore has been overharvested and become endangered. To avoid participating in the appropriation and exploitation of ancestral medicine, Vick says, "If you can grow your own plants without the intention of profiting off of them, I think that's how we stay in right relationship."

4 Editors of *Scientific American*, "Biodiversity's Greatest Protectors Need Protection," *Scientific American,* October 1, 2021, www.scientificamerican.com/article/biodiversitys-greatest-protectors-need-protection/.

5 While it's still a relatively novel concept for municipalities in the so-called United States to pay a land tax, it's been done before. For instance, in June 2021 the Alameda City Council (located on unceded Lisjan Ohlone land) voted to include Shuumi Land Tax in the city's two-year budget.

Heather Rosenberg

1 Spanish colonizers established a permanent outpost on the banks of the Los Angeles River in 1781, and by the 1930s the city had significantly expanded into areas prone to flooding. The River Project explains on their eponymous website, "The conflict between the city's heedless development and the river's unpredictable nature led the region's leaders to partner with the US Army Corps of Engineers to confine all the river's waterways with cement encasing, fixing its course within a concrete straitjacket and defining a new standard for water management in the region for generations to come" (www.theriverproject.org/lariver).

2 According to the River Project, the Los Angeles River channels about 196 billion gallons of precious rainwater into the ocean each year, since it cannot infiltrate through the concrete encasing to local groundwater basins. For context, the Los Angeles Basin imports approximately 244 billion gallons of water from Northern California and the Sierras each year.

3 As of 2020 the Los Angeles Homeless Services Authority estimates that 66,000 Angelenos lack access to housing. Los Angeles's unhoused population accounts for approximately one-fifth of all unhoused folks in the so-called United States.

4 According to US Census Bureau estimates for 2018, the median annual income for white households in Los Angeles was $87,393, while the median annual income for Black households was $48,519. Median incomes for Asian, Pacific Islander, Hispanic, and Native American households fell in between. The wealth gap is even more pronounced; according to a National Asset Scorecard for Communities of Color survey conducted in 2016, white households in Los Angeles had a median net worth of $355,000, compared to a median net worth of $4,000 for Black households.

5 A yearlong investigation led by the *LA Times* and a team of environmental epidemiologists found that 3,900 people died from extreme heat in California between 2019 and 2019.

6 Designing Justice + Designing Spaces is an Oakland-based nonprofit architecture and real estate development firm with a mission to end mass incarceration and structural inequity.

7 Ben Goldfarb, "The Creative, Climate-Saving Magic of Beavers," *Yes!* magazine, June 27, 2018, www.yesmagazine.org/environment/2018/06/27/the-creative-climate-saving-magic-of-beavers.

8 The Nature Conservancy, "Removing Barriers for Healthy Rivers and Fisheries," updated April 20, 2022, www.nature.org/en-us/what-we-do/our-priorities/tackle-climate-change/climate-change-stories/removing-barriers-river-health/.

9 The Nature Conservancy, "Sustainable Rivers Program," updated March 29, 2022, www.nature.org/en-us/what-we-do/our-priorities/protect-water-and-land/land-and-water-stories/sustainable-rivers-project/.

10 Richard Conniff, "Climate Change Complicates the Whole Dam Debate," *Scientific American*, March 14, 2017, www.scientificamerican.com/article/climate-change-complicates-the-whole-dam-debate/.

11 Ayana Young, interview with Samuel Gensaw III, *For the Wild* podcast, episode 256, October 20, 2021.

Cate Mingoya-LaFortune

1 Unlike Superfund sites, which are the most severely contaminated hazardous waste sites, the Environmental Protection Agency does not currently have the authority to make polluters pay for the cleanup of brownfield sites. Instead, the EPA has made small amounts of seed money available for remediation projects and handed over program regulation to states and communities. The burden therefore falls on community members to initiate and see brownfield remediation projects through.

Didi Pershouse

1 Hasan Yilmaz et al., "Determination of Temperature Differences between Asphalt Concrete, Soil and Grass Surfaces of the City of Erzurum, Turkey," *Atmosfera* 21, no. 2 (April 2008).

2 Zero budget natural farming grew in popularity during a time when a tremendous number of farmers were dying by suicide after going into debt to corporations selling fertilizer and seed. The idea was for farmers to become independent of all corporate interests and reclaim local food sovereignty.

3 Judith D. Schwartz, *The Reindeer Chronicles: And Other Inspiring Stories of Working with Nature to Heal the Earth* (White River Junction, VT, Chelsea Green, 2020).

4 Many conservationists have misunderstood the relationship between grazers and soil health. Although animals who are confined to the same area of land for long periods of time will overgraze, grazers with access to an appropriate area of land are hugely beneficial to the landscape: they are pivotal to nutrient cycling and seed spreading, preventing major fires by keeping fuel loads low, maintaining the openness of landscapes, packing the ground, preventing dust

storms, and helping permafrost deep-freeze in the winter. Not to mention, grazers are an important food source and carry great cultural and spiritual significance for many Indigenous peoples.

5 *Gather* is a 2020 documentary film about the growing Indigenous food sovereignty movement. Maddie Oatman, "A Netflix Doc Wants to Fix Our Food System with Capitalism. 'Gather' Argues That's How It Broke," Grist, December 9, 2020, https://grist.org/food/netflix-kiss-the-ground-vs-gather/.

6 These herbivorous grazers nearly went extinct after colonizers slaughtered them en masse in an attempt to steal the land, but they are crucial to the health and climate resilience of the Great Plains, fertilizing the soil, regenerating the grasses, dispersing seeds, and increasing the presence of biodiversity (not to mention that buffalo carry spiritual and cultural significance and are an important food source for many).

7 Schwartz, *Reindeer Chronicles*, 10.

Ceci Pineda

1 A nineteen-year study conducted by University of California, Davis, compared soil carbon changes in a plot of "conventional" farmland as compared to compost-added farmland. It found that the carbon content in the compost-added soil increased by 12.6 percent over the length of the study. They concluded that compost is key to sequestering carbon in cropland soils. (Nicole E. Tautges et al., "Deep Soil Inventories Reveal That Impacts of Cover Crops and Compost on Soil Carbon Sequestration Differ in Surface and Subsurface Soils," *Global Change Biology* 25, no. 11 (2019), https://doi.org/10.1111/gcb.14762).

2 Environmental Protection Agency, "Food: Material-Specific Data," November 2021, www.epa.gov/facts-and-figures-about-materials-waste-and-recycling/food-material-specific-data.

3 Methane is a seriously potent greenhouse gas, twenty-five times more efficient at trapping heat than carbon dioxide over a 100-year period! According to the EPA, landfills are responsible for about 14.5 percent of US human-related methane emissions as of 2020.

Margo Robbins

1 Jane Palmer, "Fire as Medicine: Learning from Native American Fire Stewardship," *Eos* magazine, March 29, 2021, https://eos.org/features/fire-as-medicine-learning-from-native-american-fire-stewardship.

2 Nathan Gilles, "Wildfires Are Essential: The Forest Service Embraces a Tribal Tradition," *Yes!* magazine, April 3, 2017, www.yesmagazine.org/issue/science/2017/04/03/wildfires-are-essential-the-forest-service-embraces-a-tribal-tradition.

Doria Robinson

1 Chevron Corporation owns and operates a 2,900-acre petroleum refinery in Richmond, California, on the San Francisco Bay that has been in operation since 1902. Along with several catastrophic explosions and fires (Doria recalls one explosion that turned the sky red and ate the paint off their family car), the refinery has steadily polluted the community's air, water, and soil for decades.

2 California's cap-and-trade law was passed in 2006, and the program was launched in 2013 as a central piece of the state's strategy to reduce greenhouse gas emissions over time. Countless environmental justice groups have critiqued the market-based emissions trading system for allowing large polluters to buy offset credits instead of actually reducing their emissions, particularly in low-income communities and communities of color.

3 According to the IPCC, rising temperatures, extreme precipitation patterns, and rapid loss in biodiversity and habitat will likely lead to the escalation of vector-, food-, and water-borne disease.

4 According to the ongoing Richmond Environment and Asthma Community Health (REACH) Study conducted by UCSF, UC Berkeley, and LifeLong Medical, asthma rates in Richmond, California, are about double the state average.

5 According to data captured between 2012 and 2017 from the US Federal Reserve, Black business owners were denied bank loans at a rate twice as high as white business owners.

6 Keane Bhatt, "Dangerous History: What the Story of Black Economic Cooperation Means for Us Today," *Yes!* magazine, October 7, 2015, www.yesmagazine.org/economy/2015/10/07/dangerous-history-what-the-story-of-black-economic-development-means-for-us-today.

7 This quote was shared in the *Land Rematriation and Capital in the Commons* presentation by Lex Barlowe in Movement Generation's Fall 2021 ReKINdle series.

Cassia Herron

1 According to agricultural economics professor Dr. William Snell, tobacco was the number one crop grown in Kentucky through the 1900s, with virtually every farmer growing at least some tobacco. However, due to a drop-off in demand, increased regulations of the tobacco industry, and an act known as the Tobacco Buyout, most of Kentucky's tobacco farmers no longer grow the crop.

2 Kentucky was famously resistant to the EPA's Clean Power Plan, which aimed to reduce greenhouse gas emissions from power plants and set emission targets for each state. Senator Mitch McConnell called on governors to boycott the federal rule, Kentucky's lawmakers passed a bill to exempt the state from submitting a plan to meet proposed air regulations, and the state sued the EPA over the rule.

3 Since 1998 Kentucky has received over $2.5 billion in settlement funds from tobacco manufacturers under the Master Settlement Agreement.

Marta Ceroni, PhD

1 The genuine progress indicator (GPI) framework is an indicator alternative to gross domestic product (GDP) meant to monitor positive and negative environmental, social, and economic metrics. The framework aims to shift focus beyond economic growth to assess more meaningful and holistic measures of progress.

2 Scientists have agreed that 1.5 degrees Celsius, or 2.7 degrees Fahrenheit, is the warming threshold that the planet cannot exceed if we hope to avoid the most

severe climate disruptions. Earth has already heated up 1.1 degrees Celsius since the preindustrial era.

3 IPCC, *Climate Change 2022: Mitigation of Climate Change. Contribution of Working Group III to the Sixth Assessment Report of the Intergovernmental Panel on Climate Change* (Cambridge, UK: Cambridge University Press, 2021).

4 According to 2021 data from Greenberg and Bloomberg, ExxonMobil spent an estimated 0.16 percent, or a fraction of a percentage point, of their total capital expenditures on low-carbon energy technology in 2021. ExxonMobil spent more than double that amount on executive pay.

5 For context, the Biden administration approved 3,557 permits for oil and gas drilling on public lands in its first year in office. According to the Environmental and Energy Study Institute, conservative estimates put United States direct subsidies to the fossil fuel industry at about $20 billion per year.

6 Indigenous Environmental Network and Oil Change International, *Indigenous Resistance against Carbon*, August 2021, www.ienearth.org/wp-content/uploads/2021/09/Indigenous-Resistance-Against-Carbon-2021.pdf.

7 Mary Heglar helped launch the greentrolling movement on October 22, 2019, when BP tweeted, "The first step to reducing your emissions is to know where you stand. Find out your #carbonfootprint with our new calculator & share your pledge today!" Mary iconically responded, "Bitch what's yours???" (https://twitter.com/maryheglar/status/1186979003216859136).

8 Emily Atkin, "Drag Them: The Climate Case for Calling Out Fossil Fuel Companies Online," *Heated* newsletter, December 10, 2020, https://heated.world/p/drag-them.

Crystal Huang

1 Investor-owned utility companies largely neglected rural communities. As a result, the Rural Electrification Administration was created in 1935 to provide financing for rural electric cooperatives and ensure that rural communities could access electricity, too. Today the US still has over 800 rural electric co-ops serving about 14 percent of Americans and over 80 percent of the country's landmass.

2 The Energy Democracy Project, "A People's History of Utilities," 2021, https://emeraldcities.org/wp-content/uploads/2021/11/FINAL-Peoples-History-of-Utilities-1.pdf.

3 Aviva Chomsky, "Energy Is a Human Right," *Yes!* magazine, April 14, 2022, www.yesmagazine.org/environment/2022/04/14/energy-is-a-human-right.

Moji Igun

1 The Nap Ministry can be found on Instagram @thenapministry. Tricia Hersey also published a book, *Rest Is Resistance: A Manifesto*, in October 2022.

Deseree Fontenot

1 Queer and trans communities have a 120 percent higher risk of experiencing some form of homelessness, and the statistics are even more pronounced for BIPOC queer and trans folks. Matthew H. Morton et al., "Prevalence and Correlates of Youth Homelessness in the United States," *Journal of Adolescent Health* 62, no. 1 (January 2018): 14–21, https://doi.org/10.1016/j.jadohealth.2017.10.006.

2 Police presence often spikes during and after disasters as police are often the first to be called upon to manage evacuations, facilitate search and rescue operations, and maintain order. Communities that already experience disproportionate police brutality, such as queer and trans people and communities of color, therefore often experience spikes in police violence, particularly as law enforcement suspends basic human rights in the name of extraordinary circumstances and patrols for so-called looting.

3 According to data presented by the Prison Policy Initiative, queer and trans people are arrested, incarcerated, and subjected to community supervision and particularly inhumane treatment at significantly higher rates than straight and cisgender people.

4 Lee Edelman's *No Future: Queer Theory and the Death Drive*, published in 2004, asserted that queer folks shouldn't order their lives aimed toward the future. He argued that queer folks should embrace their role as societal archenemies and spurn politics, normative society, and the heteronormative timeline of marriage,

reproduction, and raising children. José Esteban-Muñoz's *Cruising Utopia: The Then and There of Queer Futurity*, published in 2009, agreed with Edelman that queer folks should reject the requirements of heteronormative temporality, but asserted that they *should* pursue futurity, and specifically a queer utopian futurity shaped by robust queer communities of eros and resistance.

Jacqueline Thanh

1 After residents of Village de L'Est, also known as Versailles, returned to their neighborhood and began rebuilding in the wake of Hurricane Katrina's destruction, it was announced that a 100-acre landfill would be built nearby to take in a portion of the city's smashed homes and toxic hurricane debris—without even conducting an environmental impact study or holding a public hearing (which was only possible due to the emergency powers granted to the mayor in the wake of a major hurricane). Community members protested and waged legal battles at the state and federal level until the landfill project was shut down.

2 New Orleans East, one of the lowest parts of the city at zero feet above sea level, is considered one of the areas most vulnerable to climate change in the region.

3 After the Fall of Saigon in 1975, a tight-knit group of Vietnamese refugees, who had already fled rural North Vietnam decades earlier to escape communist persecution, came to New Orleans through the Catholic Church's refugee-resettlement program and were placed in the New Orleans East public housing project known as the Versailles Arms Apartments.

4 In New Orleans, white flight—the large-scale migration of white people from areas becoming more racially or ethnoculturally diverse—was particularly pronounced in the 1950s and 1960s after the US Supreme Court ruled that racial segregation of public schools was unconstitutional.

5 After the Vietnamese refugees (many who came from farming villages) settled in eastern New Orleans, a semitropical climate not too different from Vietnam, many decided to grow their own food, and a patchwork of home gardens and urban farms developed across the neighborhood.

Janelle St. John

1 Jennifer Steinhauer, "Victory Gardens Were More about Solidarity Than Survival," *New York Times Magazine*, July 15, 2020, www.nytimes.com/2020/07/15/magazine/victory-gardens-world-war-II.html.

2 Leah Penniman, *Farming While Black* (White River Junction, VT: Chelsea Green, 2018).

3 Shea Swenson, "The Untapped Potential for Urban Agriculture in Detroit," *Modern Farmer*, April 8, 2022, https://modernfarmer.com/2022/04/urban-farms-in-detroit/.

4 David Karp, "Most of America's Fruit Is Now Imported. Is That a Bad Thing?," *New York Times*, March 13, 2018, www.nytimes.com/2018/03/13/dining/fruit-vegetables-imports.html.

5 United Nations Department of Economic and Social Affairs, "2018 Revision of World Urbanization Prospects," May 16, 2018, https://www.un.org/en/desa/2018-revision-world-urbanization-prospects.

Miriam Belblidia

1 Project South is an Atlanta-based nonprofit rooted in the legacy of the Southern Freedom Movement and committed to growing community power, regional power, and grassroots leadership. They are one of the founding organizations of the Southern Movement Assembly, a movement alliance of grassroots organizations across the South committed to growing bottom-up power and building infrastructure for long-term liberation.

2 One of the strategies of the Southern Movement Assembly is building physical infrastructure across the region where people can enter the movement, gather and strategize, find sanctuary, and gather resources to rapidly distribute during and after crises.

3 Social aid and pleasure clubs developed in the late 1700s for Black folks to pool resources for healthcare costs and funeral expenses and to host social events. Many social aid and pleasure clubs in so-called New Orleans still exist

today, and though their roles have evolved, they're still a major source of cultural pride and civic engagement.

4 Dean Spade, *Mutual Aid* (New York: Verso, 2020), 1, 7.

5 Spade, *Mutual Aid*, 2, 39.

6 Spade, *Mutual Aid*, 45.

Lil Milagro Henriquez

1 Aisling Irwin, "No PhDs Needed: How Citizen Science Is Transforming Research," *Nature*, October 23, 2018, www.nature.com/articles/d41586-018-07106-5.

Amee Raval

1 Utility companies like PG&E, which serves the northern two-thirds of California, began instituting *public safety power shutoffs* during extreme weather, such as especially hot and windy days, to minimize the risk of sparking a fire with one of their power lines (and to minimize possible liability). The unpopular (and dangerous, particularly for folks who rely on power to live) Band-Aid solution to wildfire risk demonstrates the importance of comprehensive energy system reform and developing new energy systems altogether.

2 Amee Raval and the APEN team, "Mapping Resilience: A Blueprint for Thriving in the Face of Climate Disasters," Asian Pacific Environmental Network, June 2019, https://apen4ej.org/mapping-resilience/.

3 As Amee Raval mentions, the concept of *vulnerability* is loaded. The term can be useful in this context to better understand who will be most impacted by climate risk. But it often erases the strength, resilience, and solutions of so-called vulnerable communities. Vulnerability should also be appropriately contextualized: it is not an inherent characteristic or condition of groups of people; rather it is a circumstance or consequence of historic and systemic marginalization, oppression, and exclusion from opportunity. Don't be afraid to challenge this language and choose or create terms of your own.

4 Kendra Pierre-Louis explained in her article "There's Actually No Such Thing as a Natural Disaster," published in *Popular Science* in October 2017, "Most of what we call natural disasters (tornadoes, droughts, hurricanes) are indeed natural, though human contributions may increase their likelihood or intensity. But they aren't disasters—they're hazards. If a hurricane slams into land where no one lives, it isn't a disaster; it's weather. A disaster is when a natural hazard meets a human population. And often, that intersection is far from natural" (https://www.popsci.com/no-such-thing-as-natural-disaster/).

Marcie Roth

1 People with disabilities are two to four times more likely to be injured or killed in a natural disaster. Paul Timmons, "Disaster Preparedness and Response: The Special Needs of Older Americans," Statement for the Record, Special Committee on Aging, US Senate, September 20, 2017, www.govinfo.gov/content/pkg/CHRG-115shrg30022/html/CHRG-115shrg30022.htm.

2 World Economic Forum, "The Valuable 500—Closing the Disability Inclusion Gap," www.weforum.org/projects/closing-the-disability-inclusion-gap.

3 US Centers for Disease Control and Prevention, "Disability Impacts All of Us," September 16, 2020, www.cdc.gov/ncbddd/disabilityandhealth/infographic-disability-impacts-all.html.

4 This slogan of the disability rights movement came into use by South African disability activists William Rowland and Michael Masutha to mean that no policy should be decided without the full and direct participation of members of the groups affected by that policy.

Eileen V. Quigley

1 In 1946 the US government decided to conduct nuclear weapon tests on Bikini Atoll, a ring of small coral islands that are part of the Marshall Islands, then under US control. The US military demanded that the residents of Bikini Atoll leave, and subsequently detonated twenty-three nuclear devices between 1946

and 1958. The islands were so thoroughly contaminated by radiation that they are still not safe to return to.

2 Coal combustion emissions are linked to damage of the respiratory, cardiovascular, and nervous systems and contribute to four of the top five leading causes of death in the so-called United States, including heart disease, cancer, stroke, and chronic respiratory disease.

3 Lowell Ungar and Steven Nadel, "Halfway There: Energy Efficiency Can Cut Energy Use and Greenhouse Gas Emissions in Half by 2050," ACEE, September 18, 2019, www.aceee.org/research-report/u1907.

4 Kate Zerrenner, "Why Energy Efficiency Is Key to Reducing Climate Change Risks," Triple Pundit, January 3, 2020, www.triplepundit.com/story/2020/why-energy-efficiency-key-reducing-climate-change-risks/86086.

5 Andrew Reimers, "Making Electricity Consumes a Lot of Water—What's the Best Way to Fix That?," *Scientific American*, May 17, 2018, https://blogs.scientificamerican.com/plugged-in/making-electricity-consumes-a-lot-of-water-whats-the-best-way-to-fix-that/.

6 E2 and E4TheFuture, *Energy Efficiency Jobs in America*, October 2021, https://e4thefuture.org/wp-content/uploads/2021/10/Energy-Efficiency-Jobs_2021_All-States.pdf.

7 Jason Hickel, *Less Is More* (London: Penguin Random House UK, 2021), 152.

Natalie Hernandez

1 Environmental Protection Agency, "Keeping Your Cool: How Communities Can Reduce the Heat Island Effect," November 2014, www.epa.gov/sites/default/files/2016-09/documents/heat_island_4-page_brochure_508_120413.pdf.

2 Jeremy S. Hoffman, Vivek Shandas, and Nicholas Pendleton, "The Effects of Historical Housing Policies on Resident Exposure to Intra-Urban Heat: A Study of 108 US Urban Areas," *Climate* 8, no. 1 (January 2020), https://doi.org/10.3390/cli8010012.

3 Rebecca Leber, "How to Redesign Cities to Withstand Heat Waves," Vox, June 30, 2021, www.vox.com/22557563/how-to-redesign-cities-for-heat-waves-climate-change.

4 Philip Oldfied, "What Would a Heat-Proof City Look Like?," *Guardian*, August 15, 2018, www.theguardian.com/cities/2018/aug/15/what-heat-proof-city-look-like.

Chief Shirell Parfait-Dardar

1 As sea levels rise and saltwater is pushed into swampy, freshwater-dependent ecosystems, low-lying, coastal trees become bleached, bare, and blackened, eventually dying altogether. They are often referred to as "ghost forests" for their distinct and ominous appearance.

2 As part of the criteria for acknowledgment as a federally recognized Indian tribe under 25 CFR §83.11(c), "the petitioner [must demonstrate] that it comprises a distinct community and [has] existed as a community from 1900 until the present," and one of the pieces of evidence that petitioners may use to sufficiently demonstrate distinct community and political authority is that "more than 50 percent of the members reside in a geographical area exclusively or almost exclusively composed of members of the entity, and the balance of the entity maintains consistent interaction with some members residing in that area." Without federal acknowledgment, tribes are ineligible for federal assistance from the Federal Emergency Management Agency and the Bureau of Indian Affairs.

3 The nearby Isle de Jean Charles, the ancestral homelands of the Isle de Jean Charles Band of Biloxi-Chitimacha-Choctaw, has lost 98 percent of its landmass due to rising waters since 1955, when tracking began. In 2016 the residents of Isle de Jean Charles became the first federally funded climate migrants in the continental so-called United States when Louisiana's Office of Community Development received $48 million from the Department of Housing and Urban Development to resettle current and former residents of Isle de Jean Charles to a "New Isle" forty miles inland. While many residents will be relocating to the New Isle, some are staying behind, like Chief Naquin, who says that "the state's plans for the community have lost their original intent

of allowing the tribe to be in charge of the sustainable vision for their new home." Chief Shirell Parfait-Dardar critiqued the relocation process, saying, "They clearly are following a system that does not understand us or even want to, that same system of assimilation."

Eve Mosher

1 According to data from Climate Central on the *States at Risk* platform, Philadelphia is the seventeenth fastest warming city in the so-called United States.

Irfana Jetha Noorani

1 According to a University of Minnesota Law School study, "American Neighborhood Change in the 21st Century," published in 2019, Washington Navy Yard saw its population grow by 251 percent between 2000 and 2016, while its percentage of low-income residents dropped from 76.9 percent to 20.8 percent.

2 As of 2019, Black residents make up 84.1 percent of Anacostia's population, and in fact, about half of DC's Black population resides east of the Anacostia River. The area became predominantly Black in the 1950s amid school desegregation and white flight into surrounding suburbs. Anacostia subsequently experienced damaging urban renewal programs, major disinvestment in basic living necessities, highway construction through established neighborhoods, and other damaging projects and policies that contributed to concentrated poverty and environmental injustice. The neighborhood also evolved as a hub for civil rights and Black activism and community organizing.

3 The High Line in NYC and 606 in Chicago are both major infrastructure reuse projects transforming abandoned rail lines into multiuse community gathering spaces. While both projects were successful in many regards, they also spurred luxury development and runaway gentrification.

4 The 1995 Chicago heat wave was the deadliest extreme heat event in the country's history. Temperatures remained over 100°F for five days, and with unbearable humidity, the heat index reached 126°F.

5 Christine E. Nieves, "Community Is Our Best Chance," in *All We Can Save*, ed. Ayana Elizabeth Johnson and Katharine K. Wilkinson (New York: One World, 2020), 366.

6 Note that Creation Justice Ministries' webinars, guidebooks, and other resources are primarily geared toward Christian churches.

Melissa Reyes

1 Since we spoke, Melissa stepped down from her role at the Center for Story-Based Strategy to shift her focus to healing work.

2 Rob Hopkins, *From What Is to What If: Unleashing the Power of Imagination to Create the Future We Want* (White River Junction, VT: Chelsea Green, 2019), 119, 120.

Patty Berne

1 Patty and Mia's vision for disability justice soon expanded to include others like Leroy Moore, Stacey Milbern, Eli Clare, and Sebastian Margaret, who have all been instrumental in shaping the disability justice movement.

2 Leroy F. Moore Jr., a Black, disabled community activist, author, and poet, is considered a leading voice in the exploration of the intersection of ableism and racism.

3 Essex Hemphill, who lived from 1957 to 1995, was a poet and performer known for addressing the intersections of race, identity, sexuality, and HIV/AIDS in his work.

4 For disabled people to be eligible to receive Supplemental Security Income, the individual income limit as of 2022 is $841 a month (about $10,000 a year), and the countable resources limit (including cash and financial assets) for an individual cannot exceed $2,000.

5 Indeed, of those who perished in the Camp Fire, the average age was seventy-one. Many had physical disabilities, could not drive, and/or did not have access to a car. Paradise, California, was once considered a haven for retirees

with limited income, and thus the percentage of the population that was disabled and/or older than sixty-five was more than double the statewide rate.

6 Patty noted that she hates the word *inclusive* but it's useful in this context.

7 While the term *crip* has historically been used to stigmatize and oppress disabled people, many disabled people have reclaimed the term.

8 According to a Bureau of Justice Statistics survey in 2016, 38 percent of state and federal prisoners reported having at least one disability. Disabled people are significantly overrepresented in all interactions with the criminal justice system; they are more likely to be stopped, arrested, and murdered by police, and they are less likely to have access to the services needed to successfully navigate the court system.

9 In this context, to crip something means to apply a disability justice lens to it.

10 Micah Bazant is a visual artist and cultural strategist who works with social justice movements to reimagine the world. Toni Cade Bambara, who was a civil rights activist, author, professor, and filmmaker, is known for saying, "The role of the artist is to make the revolution irresistible."

Selin Nurgun

1 The Work That Reconnects circles range in format but generally consist of regular gatherings providing space to study, practice, and build community around the Work That Reconnects framework based upon the teachings of Joanna Macy.

2 Emma Goldman, an anarchist political activist and writer, shared in her autobiography *Living My Life* that when she was scolded for dancing because "it did not behoove an agitator to dance," she grew furious; "I did not believe that a Cause which stood for a beautiful ideal, for anarchism, for release and freedom from conventions and prejudice, should demand the denial of life and joy. I insisted that our Cause could not expect me to become a nun and that the movement should not be turned into a cloister. If it meant that, I did not want it."

Sekita Grant

1 Hurricane Michael devastated the Florida Panhandle in October 2018, killing at least forty-five people and damaging an estimated 60,000 homes.

2 Countless studies have documented that FEMA provides more assistance to white and wealthy individuals and communities while shortchanging low-income and communities of color. Dr. Samantha Montano explains in her 2021 book *Disasterology* that this is a feature, not a bug, in the system; the limited intervention model of the national emergency management system, which only works for the folks with the time and resources to navigate the process, was designed to uphold white elitism.

3 Valeria Lvovna Gelman and Daniely Votto, "What If Citizens Set City Budgets?," World Resources Institute, June 13, 2018, www.wri.org/insights/what-if-citizens-set-city-budgets-experiment-captivated-world-participatory-budgeting.

4 Chris Colin, "What If Public Funds Were Controlled by the Public?," *New York Times*, April 18, 2022, www.nytimes.com/2022/04/18/us/participatory-budgeting-shari-davis.html.

5 Courtney Martin, "Budgets Are Moral Documents: 5 Questions for Participatory Democracy Black Belt Shari Davis," *The Examined Family* (blog), July 17, 2020, https://courtney.substack.com/p/budgets-are-moral-documents.

Mara Ventura

1 Rebecca Solnit, "How to Survive a Disaster," Literary Hub, November 15, 2016, https://lithub.com/rebecca-solnit-how-to-survive-a-disaster/.

2 Samantha Montano, "Bootstraps," in *Disasterology: Dispatches from the Frontlines of the Climate Crisis* (New York: Park Row Books, 2021), Kindle.

3 Montano, "'A Real Catastrophe Like Katrina,'" in *Disasterology*.

Kavaangsaar Afcan

1 Pebble Mine is a gold and copper mine proposed to be constructed in the Bristol Bay headwaters, a watershed that provides vital habitat to numerous species and is home to one of the greatest wild salmon fisheries on Earth.

Bristol Bay residents, nearby tribes, fishermen, environmentalists, and allies have been organizing against the mine's construction for the past two decades, citing that the proposed project would irreparably harm the ecosystem, surrounding communities, and salmon fishery. The Environmental Protection Agency effectively vetoed the proposed Pebble Mine in February 2023.

2 According to a yearlong study conducted by the First Nations Development Institute in 2017, the average food basket (of staples like white bread, eggs, milk, apples, and coffee) in Alaska Native communities cost 2.5 times as much as the national average.

Olivia Juarez

1 The Bureau of Land Management is a federal agency that helps determine how public lands are used, whether for recreation, timber harvesting, livestock, grazing, or fossil fuel and mineral extraction.

Sona Mohnot

1 Sona defines *grass-tops* as follows: "Whereas grassroots orgs have a strong membership base and are accountable to the community members where they're located, and there's usually an organizing arm, grass-tops orgs generally lack that membership base. They're not necessarily only working in the community that they're working in. Organizations like the Greenlining Institute do some local work, but the focus is at the state level, so it's really important for the Greenlining Institute to partner with grassroots organizations on the ground."

2 According to data from the 2019 Survey of Consumer Finances, the median white family in the so-called United States has eight times the wealth of the median Black family and five times the wealth of the median Hispanic family (and other racialized groups fall somewhere in between).

3 Indeed, a 2022 study conducted by UC Berkeley researchers of more than 200 cities across the nation found a strong link between formerly redlined neighborhoods and higher levels of air pollution: residents of formerly redlined neighborhoods breathe an average of 56 percent more of the freeway pollutant nitrogen dioxide and suffer from higher levels of the sooty

particle PM 2.5. Both pollutants are linked to higher rates of asthma and heart disease.

4 Find Amee Raval's essay on p. 173.

Kailea Frederick

1 Monsanto has since pleaded guilty to thirty environmental crimes related to illegally spraying banned pesticides and storing chemicals considered an acute hazardous waste on Molokai and Maui.

2 For nearly a decade, Indigenous Hawaiians have been organizing to prevent the Thirty Meter Telescope from being constructed on Mauna Kea, which is one of the most sacred sites, if not *the* most sacred, in Hawaiian culture.

3 Kailea and I spoke in the winter of 2021. Her term on Petaluma's Climate Commission has since ended, and she presently splits her time between Loam and NDN Collective's advancement team.

Dominique Thomas

1 It's important to note that a tenant can be evicted and banned from all New York City Housing Authority housing even if they are not convicted of an offense—a mere arrest is enough.

2 According to a NYC Health report published in September 2021, in high-poverty neighborhoods like Hunts Point, Mott Haven, and Highbridge, which are exposed to tremendous pollution from highways, distribution centers, commercial facilities, solid waste transfer stations, power plants, incinerators, and MTA bus depots, the rates of asthma-related emergency department visits among children are nearly twenty times higher than the rates for children residing in nearby Bayside and Little Neck, low-poverty neighborhoods in Queens.

3 Indeed, a 2018 report from Green 2.0 revealed that among the forty largest environmental nongovernmental organizations in the so-called United States, 80 percent of the staff is white.

4 Dominique and I spoke in April of 2021. She no longer works as the training manager of Climate Advocacy Lab; she presently serves as a partner and field building director for climate startup Industrious Labs.

INDEX

D

E

F

G

H

I

N

O

P

T

U

V

W

Y

Z

ABOUT THE AUTHOR

Kylie Flanagan is a climate communicator and the executive director of a small, climate justice–focused foundation. Originally from Miwok lands in the California Bay Area, she currently resides on Munsee Lenape lands in New York City. She graduated Phi Beta Kappa from Dartmouth College and received a master's in sustainability solutions from Presidio Graduate School. She has dabbled in goat midwifery, cheesemaking, tiny house architecture and construction, supper club hosting, edible landscaping, and sustainable business consulting, always driven by a desire to make the world more delicious, beautiful, joyous, and just. *Climate Resilience* is her first book.

About North Atlantic Books

North Atlantic Books (NAB) is an independent, nonprofit publisher committed to a bold exploration of the relationships between mind, body, spirit, and nature. Founded in 1974, NAB aims to nurture a holistic view of the arts, sciences, humanities, and healing. To make a donation or to learn more about our books, authors, events, and newsletter, please visit www.northatlanticbooks.com.